Deep-Rooted Wisdom

Deep-Rooted Wisdom

Skills and Stories from Generations of Gardeners

AUGUSTUS JENKINS FARMER

Timber Press
Portland | London

Published in 2014 by Timber Press, Inc.

The Haseltine Building   6a Lonsdale Road
133 S.W. Second Avenue, Suite 450   London NW6 6RD
Portland, Oregon 97204-3527   timberpress.co.uk
timberpress.com

Printed in China
Book design by Breanna Goodrow

Library of Congress Cataloging-in-Publication Data
Farmer, Augustus Jenkins.
  Deep-rooted wisdom : skills and stories from generations of gardeners /
by Augustus Jenkins Farmer. – 1st edition.
      pages cm
  Other title: Skills and stories from generations of gardeners
  Includes index.
  ISBN 978-1-60469-452-9
  1. Gardening. 2. Traditional farming. I. Title.
II. Title: Skills and stories from generations
of gardeners.
  SB454.F27 2014
  635–dc23

                    2013033871
A catalog record for this book is also available from the British Library.

For my parents,
who saw the magic in a run-down farm,
moved us in, taught us to dream,
cultivate, and, above all, share the spirit
of that special piece of earth.

Contents

Stacking Up

*Growing Plants for Food,
Construction, Flowers, Teaching,
and Connections*

Building Fertile Soils

*Encouraging a Healthy
Web of Life*

Stop the Tilling Cycle

*Harnessing the Natural Powers of
Worms and Mushrooms*

Watering by Hand

*Using the Essential Skill of
Observation to Keep Plants Hydrated*

Rooting in the Ground

*Working with
Pass-Along Plants*

Saving Seeds

*Treasuring Heirlooms for Genetics
and Nutrients*

Foreword

EVERY SCHOOL CHILD CAN name the four basic flavors, ticking off on their fingers the sweet-sour-bitter-salty mantra. But just as there's a fifth finger, there's another flavor that most of us overlook or simply weren't taught because it can't be easily described, any more than we can describe a cat's purr: It's *umami*, or savory. Think mushrooms, olive oil, oysters, or avocados. Their flavors don't fit neatly into traditional taste tests, so we just say, "good." Same with a cat's purr.

Likewise, many gardening books, like formal education, are mostly derivative, transferring old bones from one pile to another while teaching us methods of coloring inside the lines. This is important for goal-oriented horticulturists who are all about results; soil testing, pruning just so, planting in rows, special soil mixes, and all those other tools and techniques make sense from a purely productive perspective.

But our right brain urges us to slow down occasionally, to leave efficiency in mid-stroke and savor little unexpected experiences. There is magic in the everyday, and our physical senses are ready to receive. Once you smell fresh-cut basil, just seeing a photo of it will conjure the fragrance in our mind. We need to feel the hot sun on the back of our hands, or raise our arms a bit to let a sudden breeze chill the sweat under our shirts, smile when a dragonfly lands on our tomato stake, taste the tangy sourness of a clover flower stalk, and pay attention to a wind chime as it interprets an otherwise silent breeze into language our eyes and ears can understand.

This is what Augustus Jenkins Farmer—Jenks, as I know him—is all about, and more. This book shares his take on both the left-brain basics of how we garden—the quintessential tools and techniques—as well as the intangibles of why we love what we do.

As he connects us with fellow "dirt" gardeners, the kinds of folks who, depending on where they garden, make their own fig preserves or rhubarb pie, and who learn as they go, we are introduced to the spirit of "making do" with what we have on hand, finding grace and whimsy in the everyday.

The anecdotes and people of this book are like clumps of orange daylilies, which are found in every corner of the gardening world yet are generally not sold anywhere; every one you see was put there by a sharing gardener, or escaped on its own to find better dirt across the road.

Over the many years I have known and admired Jenks and his partner, Tom Hall, I have seen their yearning for simple answers to the mysteries of the garden. They are not anti-innovation Luddites, but men striving to enjoy their lives and gardens, and share with others.

Here Jenks has finally put down on paper a few of his fondest memories and connections between plants and people. We all have these stories—my favorite is when beloved southern writer Eudora Welty once told me about her mother quitting a garden club because they stopped swapping plants at their meetings. A real gardener will understand completely.

Don't just pick this up as a gardening guide; as you pick this man's amazing brain, and read between the lines for glimpses of his wry humor, feel his loving heart as well.

—FELDER RUSHING

Acknowledgments

MANY GARDENERS SEEM AS compelled to share their joy and wisdom as they are their plants. Many of my teachers taught themselves and discovered the secrets of gardening themselves, but wanted to pass it all on by molding dirt and youngsters—by sharing—without any idea of what might grow out of it all. Thank you to all the teachers and contributors in this book for sharing your ideas—sometimes pounding them into my head—and for trusting me with your friendship, your narratives, and your plants.

Most teachers and gardeners have big support networks. I've been able to call on so many people from different parts of my life to help bring this book together. Thank you to my family, Tom Hall, Gloria Farmer, Hunter Desportes, Bob Waites, and the Raintrees. To people who have read, commented, and molded, including Sharon Altman, Genice Jones, Patrick Butler, Jean Ann Van Krevelen, Bonnie Thurmond, Dick Birr, Patty Barrett, Judy Caldwell, Pam Beck, Charles Lesser, Roger Swain, and Felder Rushing.

Gardeners everywhere, even some I didn't know, introduced me to people they thought I needed to learn from, including Angie Smitman, Lindsey Kerr, Bob Fuller, Marcus Thompson, Chris Gajewicz, and Cindy Vann.

Great thanks to these photographers for sharing: Chris Brinkinshaw, Patrick Butler, Andy Cabe, Hunter Desportes, Tom Hall, Will Hooker, Jim Martin, Mark Lee Photography, Sharon Wilson of Scribbletime Photography, David Schelling Photography, and Virginia Weiler.

Finally, I'm overwhelmed by the time and gentle guidance that Jamie Hendrick contributed to this project. He became my companion on road trips to interview gardeners and give presentations, and my mentor in writing. Through his work, Jamie's taught thousands of teenagers to love words. I can't imagine the patience that's required to do that or to wrangle this mass of ideas and roots into something that might teach and inspire others.

Introduction

IF YOU'RE LIKE ME, you can look at a situation, a style, a practice, and see beyond groupthink and rote activity to find simple, effective, and fun ways of doing things. People might call us pot stirrers—I don't know how *you* caught it, but I was born this way. Looking beyond convention and questioning anything and everything has led me to create gardens stripped bare of the purchased and pretentious, overflowing with (to quote Joan Baez) the "special, miraculous, unrepeatable, fragile, fearful, tender, lost, sparkling ruby emerald jewel, rainbow splendor" stuff of life.

It was in the third grade that I became a gardener. My parents set me up in a paradise—an old farm full of rocks, boards, and scrap metal; an ancient yard filled with crinum and red spider lilies, right next to a magnolia forest carpeted with ferns. Those were my supplies. I was free to dig, transplant, and make little walkways and rooms in any way I wanted. I pruned the vitex tree, divided giant squill, and with old paint, my daddy and I painted a 12-foot sunflower on the barn wall that my garden backed up to. It's all still there thirty-five years later. It wasn't even until my second year in college that I realized that other people were building their gardens from store-bought plants and stones. Through my little garden, I'd sow seeds I found in parks, cut flowers, and try to figure out how the people who'd gardened before me might have used arrowheads, china shards, or plow parts that I unearthed. I even left little caches of toys and coins for those who'd come after me. At some point, punk rock and other teenage distractions came along and kept my attention for a while, until in college one of my cousins looked past my yellow mohawk and at my collection of acorns in the dorm fridge and said, "You ought to be in horticulture."

The idea had a lot of merit, but I found that the horticulture of academics and industry, with its focus on reductionist science and the marketing

The magical garden that I explored and lived in as a child, only to come back to later. My great, great grandfather pulled the cypress trees from the swamps, split them, and made this fence. We have cared for it ever since.

of newer and newer products, differed distinctly from the fuzzy, magical gardening of my youth. It still does. Nonetheless, I studied, was inspired by, and learned from horticultural scientists for almost eight years of my college education. For thirty years since, I've tried to mix up a holistic broth of magic, art, science, and industry with plants, worms, and dirt. I've watched lots of new products, new advertising, and new rules come along regularly, each creating unneeded complexities that only serve to intimidate future gardeners and obscure the joy of gardening.

Recently, an 84-year-old friend and I were tying plastic grocery bags around his garden—he thinks their rustling scares deer—when he said to me, "It's like all those medicines they want me to be on: they give me one, then another, and one causes the other to go wrong, and then yet another to try to fix the problem that the one caused." Soon, you're wondering if all this has fixed anything. Of course, some improvements, prescriptions, and technology do make things better. But they also need constant filtering, monitoring, and editing. Much the same can be said about gardening and the myriad prescriptions that "experts" and companies offer in literature and the marketplace. When I walk through the product-lined shelves of some big-box store's lawn and garden section, I inevitably find myself asking the same questions: How did we get here? How did we go from cuttings and manure and seeds and fun to this commercial maze? What happened to trading plants between friends and strangers? To letting things go to seed so they'd come up next year? To making a garden with things that you find lying around the neighborhood? To watering with a hose? Who ever came up with the idea of motorized garden sprayers, anyway?

This book is intended to help us—professionals and amateurs alike—escape some of these distractions; to find our way back to successful, joyful, simple gardening. Our grandparents gardened like farmers, applying their skills and know-how—to borrow one of my nana's terms, they used their make-do. Today, however, we've lost a great deal of that; there are certain essential old gardening habits that aren't even part of our discussion. We've even changed our vocabulary, losing some fascinating, useful, or simply colorful old gardening terms. Sometimes, as professionals, as lifelong gardeners, we don't take the time to consider how we've arrived at modern gardening. Sometimes we wrap skills up into blind routines, a few words, and forget the nuances of the basics. Even simple instructions can be completely miscomprehended. A "water that in, please" can lead to disappointment when you later find a tiny sprinkling of droplets around a wilted plant, or a seedling smothered in a flood of roiling watery mulch. Even in the seemingly simple act of watering, there are centuries of wisdom. That wisdom used to be passed on seamlessly from older to younger gardeners, but that kind of knowledge transfer is being lost. I cherish the people who inspired me, and I cherish more than anything the chance to be a garden mentor myself, to pass on and share stories from the garden that help us understand how we got to where we are—not just as gardeners, but as people; as stewards of the earth and all of its inhabitants.

Growing up on our little farm, every Sunday after church we'd gather for dinner. Every aunt, uncle, grandparent, and cousin sat around the table and told stories that often ended with us laughing through tears at the absurdity of life. I can still hear them now. This book is my attempt to join into that Sunday dinner storytelling. I want to honor the people who've taught me and share their lessons, their charms, and their gardens with you. Gardening the way these rural people did—and still do—often seemed to clash with the horticulture industry. But digging deeper, we find that academia, science, and industry actually validate these long-used practices. We can merge the old with the new, finding new joy in our gardens. During the short span of my career, an entire new science of soil microbiology has emerged; we understand more how essential fungi are to growing plants, and we have new questions that no one thought to ask before. There is always exciting new research in horticulture, creating new paths that can lead us out of the maze. In this book, I'll describe how on that same little family farm, we try to do this today through our garden design firm and field nursery.

Each chapter is divided into three sections. In the first section, I'll introduce an old skill or idea, something that used to be common in gardening, and take a brief look at how it's changed, how it's gotten complicated over the past few decades. In the second part, I'll introduce people who've taught me about the old ways of doing things. I'll share the wisdom of my mentors and other teachers. While most are older, there are many young gardeners who inspire me, too. And most are southern. But like lots of farm boys, I ran hard before I came back; I found teachers across the globe. In the third section, I'll combine that wisdom with commonsense ways that I have adapted and updated these lessons for modern gardening. Sometimes I do that on our farm, and sometimes in the gardens I design for my clients. In all our work, we arrive at combinations that can be artful and soulful and sometimes even absurd.

The most important lesson that I can pass on in this book has been taught to me time and time again by the gardeners of all ages that I've met along the way: take care of the earth and help others to love and respect the beautiful rainbow of the stuff of life.

Stacking Up

Growing Plants for Food, Construction, Flowers, Teaching, and Connections

O VER THE PAST DECADES, our plant palette has changed in unexpected ways. Of course, taste and trends are always changing, and plant explorers and breeders are always introducing new plants into nurseries—along with garden and landscape designers and decorators, among many others, they become the tastemakers, slowly altering which plants we can get hold of. Lots of these trends are obvious, but like the changes in the life and tilth of our soil, some can be slow and more subtle. One slow but enormous change is our shift from a rural to an urban society. As a result, we've gradually lost our connection to the land and to a time when plants were many things: our pantry, hardware store, art supplies, and medicine cabinet. Reconnecting with that time and passing down that connection should be easy; it just requires a shift in the way we think about our gardens and the purposes they serve, starting with choosing the right plants.

My mother, Gloria, in the garden where I grew up. In this little garden, on this little farm, a clump of bamboo has provided for hundreds of structural needs, including holding up the clothesline.

Not too long ago, lots of suburbanites had ties to the country. Most of us were just a generation removed, with grandparents or relatives still living and gardening on small farms. Those farmyards were intriguing jumbles of all different kinds of plants, swings, little outbuildings, and chickens. But, for many, that farm life—the plants and jumbled landscapes—came to represent hard work, repression, and poverty, while planned landscapes in the suburbs represented ease, simplicity, and pleasure. Plants became perfunctory, required by contract or expectations. Intentionally designed, aesthetic gardens were becoming more common—a luxury of people with lots of free time or cheap labor.

Standardization in all fields led us to become less reliant on our plants for use as medicine, food, construction materials, and art supplies. Who wants to go to the trouble of mixing their own paints, a process that involves growing and crushing and mixing with who knows what? Who wants to dye their own cloth—or, for that matter, grow their own flax and weave their own shirt? Our need for gardens teeming with multipurpose plants dissipated. Gardening, like so many other aspects of our modern world, became more and more about buying things.

Like many of my colleagues, I've made a living and excelled at building big, pretty gardens for my clients—a friend calls this "tarting it up." But I've always seen myself more like a missionary than a salesman. I'm driven to share the spirit of the earth and the satisfaction derived from bringing forth nutrients and joy from the dirt with as many people as possible. I knew in college that I wanted to work in botanical gardens: gardens of stewardship and service. I wanted to work in public settings where researching new ways of gardening and then interpreting and teaching those methods had equal importance. There I found teams of gardeners, volunteers, and visitors who were connected by a shared vision. I've come to feel that, together, we need to take landscapes and gardens back, from being places of consumption, places that use up the earth's resources and fill up landfills, to places that produce and supply things for our lives. Again, it starts with the plants.

What do you think of when you hear the term "cottage garden"? Perhaps you imagine an English garden in the country, with little white fences and flowers spilling through, an apple tree, and Peter Rabbit rummaging in the carrot patch. That romantic and idyllic picture, which arose during the turn-of-the-century Arts and Crafts movement, came to prominence through the work of renowned garden designers Gertrude Jeckyl and William Robinson, who appreciated and mimicked the simple beauty of working class gardens. But cottage gardens are more than that and are utilized all over the world by people who have no time for studying garden design or seeking out exotic plants from specialty nurseries. These are gardens that are planted purposefully and quickly around other useful things: sheds, slaughter tables, cauldrons, pumps, discarded tables, and stump stools.

Today, they happen on city rooftops and even in fast food parking lots. As old-style cottage gardens were built around farm things, new ones are built around fire escapes, air conditioners, and parking lots. There's a Chinese restaurant near my house where the old driveway and storage shed and everything in between is filled with kohlrabi and bitter squash. I've worked on a city project converting a fire escape, basement hatch, and three parking spaces into a courtyard garden for an obsessed office manager. The point is, these gardens happen wherever there is available dirt and in places that were not originally intended or designed to be gardens.

But it's the little country gardens that most inspire me. They stimulate memories and tell magical stories of their people—they call to me. Upon seeing them, I've knocked on gates and doors from Mississippi to Madagascar, finding great plants, creative ideas, and generous people. I love rambling through them to find a new plant or an old trick, or just a great story. What motivates Mr. Willie to let English peas climb up his okra stalks? How big is the love in Hazel that she plants an empty lot with sunflowers and puts up a sign that says, "Free, pick your own"? Those little gardens still happen today, here and all over the world. And a basic principle of each is that the plants are so much more than perfunctory. Plants *make* the garden and serve many purposes, not least of which are providing food, structure, and mulch, materials and inspiration for art, and satisfaction of the soul.

In this urban cottage garden, the homeowners found space around existing walks and porches to grow the plants they treasure.

The Teachers

RICHARD HAGER AND NAN CHASE

Trying to count the functions of a plant can seem like a funny concept. We can never do it completely, nor do we really have the right—it seems totally anthropocentric. But still we do it, and we love sharing obscure ways a plant

can be eaten, cooked, sewn with, or used. Two of my mentors in stacking are old-style country gardener Richard Hager and new urban gardener Nan Chase. Both use plants in a zillion ways, including as vehicles to tell stories and to remember people from other parts of life, and they've mixed all sorts of plants together. I can think of few people with kitchen gardens, camellia gardens, or butterfly gardens, but Richard and Nan have them all together in a beautiful jumble. In old-style country gardens or in new-style city gardens, people who garden spontaneously, without master plans or design, have lessons to teach about all the uses, pleasures, and functions of mixing up your plants.

In their old-style cottage garden, Richard Hager and his partner, Noel Wallace, have seemingly just thrown everything together—it looks like a garden of happenstance. Flowers weave around little sheds, doghouses, barns, clotheslines, and wood stacks in the backyard. Cherry trees, hydrangeas, and viburnums—Richard's love—fill beds in the front yard. Richard loves their individual beauty, and he shows them off in typical garden fashion, contrasting their habits and textures and placing them with complementing colors. But this is so much more than an artistically arranged pleasure garden.

Trees, shrubs, and perennials with practical uses are mixed in everywhere. While touring their garden, we pass by an apple tree, and Richard begins listing its uses, among them that apple wood is durable and great for building little fences and it can add flavor to smoked meats. In our brief conversation, Richard can list up to six uses for his apple tree, alone. Through every season, something from this tree is being used in the kitchen, the garden, the compost bin, or for repairing the chicken coop fence. Nothing from this plant goes to waste, and if I were to press him, I'm sure his uses would keep stacking up.

In permaculture, the popular design philosophy, one of the principles—stacking functions—states that everything should have multiple uses. Richard's apple tree is a source of food, pleasure, and wood, as well as a scrapbook. And we're not even touching on the more cerebral, the ecological functions for all the insects and animals that use it. Fences become trellises. Chickens become fertilizer producers. Trees save energy—and, like chickens, they eventually become energy, too. In cottage gardens, stacking

Creasy Greens: Weed or Trendy Vegetable?

Of all wintergreens, nothing grows through the coldest nights, or grows so easily inside, in a pot, as creasy greens. If you want to be a real country hillbilly, just pick creasy greens off the side of the road and sauté them with some grease. In modern horticulture, creasy is called bitter cress and considered a cool-season lawn weed—it's listed on the label of just about any broadleaf weed killer you can buy. But it does have the peppery bitterness of arugula and works well in a salad mix, and anybody can grow them even in a tiny pot or bucket. Sow in late summer, just as night temperatures begin to drop into the 70s. They'll germinate in just a week to ten days. Sow in a communal bed; you don't have to put them in rows or space them apart. Like non-heading lettuces, they can grow in bunches. Start harvesting leaves when they are the size of a spoon. Harvest only the leaves, leaving the crown and growing point so new leaves will emerge. Creasy will grow through bitter cold, and it needs little fertilizer. In late spring, with lettuce, mustard, and other spring crops, it will send up little stalks with green flower buds that really do look like broccoli (one of the new, hip names for it is wild broccoli). The buds open to yellow flowers, then go to seed. You can collect seed, or if you just leave some seed stalks, you'll have a crop next year without doing anything.

happens. Richard's tells stories. In his garden, he stacks things pretty deep.

Before he had a heart transplant, Richard's garden was a regular gathering place for artists, master gardeners, and friends who would hang out in such droves that some Saturdays the garden became a community market. And all the little old ladies from the neighborhood love Richard because he keeps the farm up just like they remember their aunts and uncles doing—his is even cuter, in fact. His version is slightly amped up and certainly stocked with more rare plants than an old working garden. It's the perfect balance of care and carelessness—and it was all done in his spare time.

I've known Richard since the 1990s, and for the longest time, I knew him only as a gardener. He was everywhere—gardening for other people, dropping off things he'd grown at the botanical garden where I worked, and speaking to garden clubs. It never dawned on me that he was anything *but* a professional gardener. Then one day, many years into our friendship, I received an invitation to a retirement party for Richard Hager, assistant principal, Ridgeview High School. Shocked, I spoke to a young guy who was gardening with me who recalled his high school days with Richard in charge: "He was a hard man. He'd always say to me, 'Get back here on school property and put out that cigarette, or I'll write you up and suspend your ass.'" Richard was an old-school teacher, and lots of good old boys in town showed up to that retirement party to show their respect for the man who whipped them into being gentlemen.

Perhaps because of his limited time, everything in Richard's garden is where it is because that is where it's most useful. There are no cleverly designed birdhouses to hide the sprinkler heads, and no hidden hoses. And there is definitely no putting fruits or veggies off to the side. Sitting at a little table surrounded by hosta and caladium in the shade of an oak, Richard can simply reach up and pick a fig. This reminds him to tell me of his childhood, where he grew up in a yard full of edible, useful plants and flowers: "We did everything outside, in the yard where we had apples, roses, tulips, rhubarb, strawberries, peaches, grapes, cherries, and lilies all mixed around the house and tool sheds"—all those are beautiful and edible plants. They were the backbone for the cottage garden. He says:

> I think of an old, green apple tree, growing right in the yard. You know, in warm climates you grow those summer apples, kind of little, mostly green, but so good for fried apple pies. We'd pick bushels, peel them and cut them in the yard, slice them, and put

Drink Your Yard

New plants come into our diets every day, many of them in our beverages. Think of all the new fruits, herbs, and sweeteners you've tried in juices or tea. Yaupon holly, a common landscape shrub, was a coffee substitute for the Spanish and other settlers on the East Coast. Native Americans used it as a ceremonial drink. Europeans recorded purging rituals of Native Americans, then gave the plant the scientific name *Ilex vomitoria*—because if you drink gallons of it, you'll vomit. Yaupon, also called cassina tea, has had various failed attempts at commercialization. In the 1970s, it was served in progressive coffee houses, and in the 1990s, it was promoted for its high antioxidant and mild caffeine content. An Asian drink made from *Crinum latifolium*, which legend says was so rare it was only served to the Vietnamese royal family, is now sold in premade tea bags. I mix mine with a little bourbon, limoncello, ginger zest, and ultra hot Blenheim ginger ale for an afternoon tea. Food and fruit production in a small yard can be a bit of work, so be creative in thinking about your teas. Producing, harvesting, and storing the leaves of plants is generally easier than dealing with fruits, and you can do it whenever you have the time.

Ilex vomiteria 'Carolina Ruby'

out hundreds of slices to dry in the sun on the tool shed roof— we'd scrub the roof really good first. They'd hang in cloth bags in the pantry. Momma used the dried apples to make little fried apple pies all winter long.

He recalls the group effort, from a house of nine, that it took to keep that garden, which they depended on for so many things. His grandmother's contribution was to start flowers, tomatoes, and peppers from seed underneath a wood stove in her bedroom. In this way, they all became farmers and gardeners.

Today it's just Richard and Noel putting in all the work, and at sixty-five, Richard is still learning from their garden all the time. He asks me to look around and name all the edible things I see in the garden. I think he has two reasons for this. First, the teacher in him wants to challenge me. Second, he's inquisitive and wants to find out if I know about eating or drinking any of his plants in ways he hasn't discovered yet. I see figs, basil, rosemary, blueberries, apples, and pecan trees. Then I mention a more obscure useful

plant, yaupon holly, from which you can make a stimulating tea. He tries to top me: "I've been eating lots of lambs quarters this year—just rinse it really well, cook it like greens, and put a drizzle of lemon on it." From Michigan to Florida, lamb's quarters is a common, bothersome weed, but Richard finds a way to put it to use. There are so many things that you can eat. You don't have to work too hard to plan an edible garden, you just have to learn about what's already there, and what you can use—even when it comes to weeds. Richard continues, "I remember eating creasy greens; in fact, I just bought seeds of it. It's a weed, but it's always around in the winter and full of micronutrients." With that, he's outforaged me; nobody eats creasy today.

While Richard's garden is a recreation of an old-style southern cottage garden, Nan Chase grows a thoroughly modern adaptation of one, in the city, on less than 2000 square feet of dirt. Nan built a new, compact house in an old textile village overlooking downtown Asheville, North Carolina. Her packed garden fits right into her neighborhood; it's the sort of garden every house in the area would have had in the 1940s.

Grapes climb the house; pruned, dwarf apple trees have clematis growing up their trunks; a tiny path leads through a pretty tangle of plants. It's so

Useful, edible, multifunctional plants can be thoughtfully mixed into modern gardens. This kiwi vine obscures a cliff and retaining wall that was built to help create a flat lawn on this hillside garden.

Nan Chase's thoroughly modern cottage garden fits right into her urban Asheville, North Carolina, neighborhood, which was built as a working class mill village. Not too many years ago, everyone would have grown food in these little yards.

packed in that Nan has to stand on the sidewalk or porch to do about half of her gardening. She's not a farmer, but she eats from the yard as much as she can and uses inspiration from the yard for her writing and photography. She wrote a book about this little space called *Eat Your Yard* and another about all the great stuff you can brew into hot and cold tea—you can drink it, too. Her garden is the perfect example of what you can do in a small space, of old cottage gardens that sprawled in new settings. And it's only five years old.

When she started, it was just a slab of clay. She doesn't try to live entirely off her plot of land, but she draws all sorts of inspiration and soulful satisfaction from it. Her garden is made not for vistas or parties or to show off her design skills, but the yard still looks absolutely outstanding, with peonies, roses, irises, chamomile, more roses, grape vines, daylilies, and all sorts of gorgeous greenery and young apples and crabapples all glistening from a recent rain. Nan loves it, as do most strangers who happen to pass by. She says:

> It just feels pretty to be out in it. The other day, a complete stranger was driving by and yelled out, "That is a beautiful yard. Just beautiful," then kept on going. Five minutes later, I was still

sweeping when a young lady walking her dog also stopped to say how beautiful the yard was. Then just now, I was out picking some salad greens and fennel, and a girl walked by and said, "This is the most amazing garden I've ever seen."

It's starting to crack me up!

The feeling of sharing beauty with strangers of all backgrounds and ages is a powerful high, and makes me so hopeful about the continuing improvements in a pretty rough neighborhood. It speaks to the transformative power of a garden far beyond the "owner's" enjoyment. It doesn't take an expensive show garden to make a difference—in fact, it just takes love.

Updates and Adaptations

Tom Hall and I are farmers (or maybe new agriculturalists, or even just fellas who have fun with flowers). We run an organic flower farm and a garden design business. Our farm looks a bit like my neighbor Mr. Frank's old farm, where I worked as a teenager, with pastures and pecan trees with clover growing underneath. We have fields of lilies and vegetables planted in rows, while the constant chatter of chickens, donkeys, bees, and dogs keeps things lively and interesting. Our health and wealth depend on the quality of our plants. Healthy plants mean healthier food, and this ensures that our customers keep coming back. This garden is our experiment, our place to try out and share old gardening methods—to show how, updated and merged with new scientific understanding, putting these methods into practice can feed the world and make a beautiful business.

I've spent most of my career building gardens for other people, both public and private. In the past, I've mostly designed and gardened using typical landscaping methods. But since I've been able to experiment on the farm, I now create gardens for people who want all earth-friendly practices, from installation to care, using what I've learned there. I use all kinds of techniques, including worms, mushrooms, compost tea, and plants that support the web of life that lives in the soil. Many of these techniques have been used for centuries, though some were not well understood until recent advances in soil sciences. On our farm and in other people's gardens, we're demonstrating that both old and new ways of gardening work, and they work even better when they're combined.

In one of my design projects, I was lucky enough to be able to spend a few amazing years working on an isolated site. Swamps, tobacco fields, and quiet, good old country people who were slightly suspicious of outsiders surrounded this farm in a remote part of South Carolina. It was to become a massive pleasure garden, yet it was a hand-hewn garden in the middle of nowhere. The mostly Mexican crew and I kept to ourselves—working, eating, and sleeping on site. Conveniently, I had recently learned Spanish, and they were eager to learn English. I quickly learned to call my coworkers

On our farm, we alternate rows of our primary crop, crinum lilies, with all sorts of useful plants. In the foreground, crinum, then purple gomphrena, okra, and pecan trees beyond—all provide joy, healthy food, and income.

cuñados—my brothers-in-law. It seemed a bit like living or working on an island, maybe near their home in Vera Cruz. Though there were local stores, they didn't carry the tools or plants that we needed. So we grew a lot from cuttings and sourced a lot of our materials from the woods. They'd grown up working with machetes and repairing fishing nets—together we made for an innovative, resourceful team.

A nearby pond was an endless resource for us: we looked to lily pads for inspiration and beauty; we cooled off by diving off the dock into the water; and we ate fish for supper. Bullrush, a tall grass-like plant that grows in the shallow water, became string for tying plants to trellises. My *cuñados* taught me, using their fishing-net sewing skills, to make hanging baskets with pine needles. We used palmetto leaves as a sled to move giant pots around. They reminded me to look at the natural world as an endless supply of raw materials. We built trellises from saplings, footbridges from tree trunks, and fences from thick muscular vines.

On the weekends, I took them on field trips—for their first elevator ride, movie, and art museum. Through it all, I watched them register wonder, disgust, and inspiration. Once, after visiting the gallery of a basket weaver, we went to my crinum farm where I had a massive pile of muscadine vine prunings. We made head-high, spherical vine balls. These mesmerizing spheres

Tom Hall and I in our crinum lily field, where most fertility comes from companion planting, microbes, and worms.

could be rolled around the yard. Later, we suspended them 20 feet high in the canopy of a whispering pine grove, where they'd quietly sway and bounce. Adult visitors would crane their necks and ask, "Do they have some purpose? Or are they just there?" Depending on my mood, I'd say they were statements on the emptiness of garden art, dancer cages, or spirit houses. Young visitors, on the other hand, just got it—they wanted to dance under them and throw things at them to make them swing.

We did, of course, also use our wild supply for serious construction needs: for trellises, fencing, bridges, and shading; for supporting and training new trees. For example, if you want a weeping tree to grow up big and tall quickly, you can put a big stake beside it and train one weeping branch upright. That branch soon becomes a trunk, and you repeat the process. Saplings and bamboo work great for this. My *cuñados* had many different words to refer to different sizes and types of bamboo, evidence of its cultural importance for them. In fact, bamboo is pretty important in most of the world. With one single, easily cut cane of bamboo, you can make a fishing pole or a tomato stake, prop up a clothesline, fence, or window unit, stick it like rebar into a wall, roast a hot dog, or make yourself a simple bowl, cup, candle, or hose pipe. Bamboo is the duct tape of the plant world. Besides, when it's alive and growing in the ground, it provides shade, privacy, erosion control, and a beautiful leaf litter that you can use as mulch on other garden beds. What's more, the edible shoots are rich in seventeen micronutrients, amino acids, and fiber.

Gardeners everywhere both love and fear bamboo. But that fear is based on stereotypes and ignorance, on limited knowledge of one type of bamboo that can, when neglected, take over other plants. Of the more than 1000 species of bamboo, there are many more true clumpers than aggressive runners. And there are ways and situations for either to add beauty, shade, vertical green, and an unlimited supply of construction materials to landscapes. I now include bamboo in every garden I build—every one. Their filigree leaves rustle; their vertical, ribbed canes mesmerize. In the right place, surrounded by asphalt or buildings, running bamboo works well—the bigger, the better. I've used giant timber bamboo in everything from small courtyards to big parking lots to provide shade and amazement. Black or gold bamboo makes for elegant container plants. For over eighteen years in my small city garden, I've grown a hedge of *Bambusa multiplex*, the true clumping bamboo (my favorite cultivars are 'Alphonse Karr', with golden canes, and 'Riviereorum', with slender, solid canes). The clumps are now just 5 feet

Using Dried Vines

Whenever I put a new trellis in a client's garden, I cut a big piece of wisteria trunk, dry it, and then wire it to the trellis post. It becomes a guide for the new vine and a way to help the new trellis feel a bit older, a bit grown in. Cut vines make great fences, ropes, swags, and plant guides. Certain vines last longer than others, depending on the water content of the wood when it's growing. Wisteria, which grows really fast and is full of water, is fun to work with but will only last one rainy summer in the South unless it's kept inside. Grape vines, pepper vines, and Virginia creeper—all slower-growing, more dense woods—will last for years. *Berchemia*, the spectacular vine of southern swamps, is also called supplejack, which, for years, I thought referenced the sinuous, snake-like ways it winds around stumps and hangs in curled swags like lianas that Tarzan might have swung on. But then a weaver friend explained to me that the name refers to the amazing flexibility of the wood, which can be used to make basket handles and embellishments. For its usefulness, sculptural stems, and for its pretty, yellow fall leaves, supplejack is a must-have, multiuse vine.

Woody vines woven to be objects of art, contemplation, or taken down and rolled around.

Berchemia twines around a cypress tree in Francis Marion Swamp in South Carolina. Because it is so flexible, weavers call it supplejack.

Dried wisteria vine, screwed onto a new trellis, acts as a guide to the young climbing rose.

31

square at the base and sheared at 15 feet tall. The hedge hides the neighbors and provides plant stakes for vegetables. There really is no need to buy those cheap, little green bamboo stakes they sell in garden centers, which actually come from tonkin bamboo, grown in mainland China. They may be easy and affordable, but there is a huge environmental cost to harvesting, cleaning, and dyeing them, packing them in plastic, and shipping them on boats and trucks to your garden.

Other trees and shrubs can also provide dried canes for staking and trellising material. In new gardens, I include cutback or coppice shrubs: woody plants that you cut to the ground in winter. By midsummer, they've grown long, straight, and useful stems. Catalpa, known around here as the bait tree, is a classic example of a tree that can be mixed into a perennial border but cut back to the ground every year. This severe pruning makes its leaves huge and pretty and its stalks really straight. In just one summer, you'll get 6-foot-tall stakes. And if you're into fishing, you'll get catalpa worms, caterpillars that make great fishing bait—I have several friends whose first jobs were to collect the worms and put them in freezers to sell later. Vitex, rose of Sharon, mulberry, and crepe myrtle can also be treated as cutback plants,

LEFT 'Tanakae' bamboo, with its green and purple leopard-skin patterns, entices people to plant it. But it is a giant plant and an aggressive spreader. This planting is in front of a commercial office building, surrounded by concrete walkways.

RIGHT Bamboo canes being dried and stored for later use at Warren Wilson College Farm in North Carolina.

STACKING UP

providing quick stems for garden staking. Japanese parasol tree can grow 15-foot-tall stalks in one summer. I cut a few every spring to have on hand for staking vegetables in the summer.

Back on our own farm, Tom and I integrate many useful plants into our crinum fields. The farm is old—it's been farmed and gardened since the 1750s, and has quite a history—so bamboo, figs, pecans, and a hundred other useful plants are established around the fields and barns. In the fields, we mix in food and herb plants. Our soil is managed to keep the microorganisms thriving. I like to call it "beyond organic." Between our lines of lilies, a row of parsley adds winter green, and in the summer, the parsley flowers attract tons of bees, flies, and wasps. By August, we collect parsley seed for cooking, and we chop and lay the withering plants down for mulch. Also, in late summer, okra provides food and can be pickled. In fall, we leave the tough stalks standing; they become trellises for February-sown sugar snaps. Then, in early spring, we cut the old okra stems.

Long, straight pieces of stems from perennials can also be useful in the garden. These lightweight, dry stems are sometimes known as haulm. Haulm is what you get when you cut and store perennial plant stalks to

LEFT Parasol tree (*Firmiana simplex*) is coppiced to show off its giant leaves. Each winter, the resulting straight stalks are cut and used for plant stakes and fences.

RIGHT Parsley, after it has gone to seed, is cut and laid over as mulch and a nutrient source.

use later. I'm a haulm hoarder. Any plant that grows a thick stem quickly is great haulm. Fern fronds were used traditionally for haulm to be applied as a light mulch and covering for frost protection. We save fennel, okra, and asparagus stems and pull them over fall veggie seedlings on the first really cold nights to act as an insulating blanket. It's the free version of what your garden store calls floating row covers. It's important to remember that you don't always need those material things on display at your garden store; you don't have to buy plastic or stakes or row covers simply to protect your seedlings from frost. As gardeners, as workers of the earth, we should be the leaders, the examples, the medium through which innovation and creativity in the use and reuse of our plants is passed on.

Thin-leaved grasses work great as mulch, too. *Spartina bakeri*, or sand cordgrass, is a favorite of mine because it's easy to harvest and lasts a long time. To keep it uniform, I grab a handful of the grass at the base and simply cut it off. Use it fresh or store it in bundles, always with the base ends together, so that all the grass is flowing the same way. Then, when you mulch with it, thread it around your plants. Having a big clump of a grass is like having your own little straw factory in the yard. If you don't have room for it, find someone else's yard to plant a grass in.

Plenty of other plants offer beautiful mulches, though plenty can be ugly. I've tried and tried to make big, curling crinum leaves into a pretty mulch. Even though they end up being ugly, I can stuff them behind and under large shrubs, under bananas, and even onto newly planned beds.

A Farm with History

We are certainly not the first to realize the beauty and potential of our spot of earth. Our soil has a history. Early people lived and cultivated crops here. At a nearby archeology site, the discovery of artifacts from as far back as 30,000 years ago suggests that people hunted along the Savannah River earlier than ever imagined. The eighteenth-century botanist and explorer John Bartram spoke of the riverside crops of the Westo Indians and the Chickasaw tribe, who took advantage of the land's rich potential. The earliest colonists knew it, too—our farm is just miles away from an English settlement that was seen on maps as early as 1685. And in the mid-1800s, Governor James Hammond, a man who dreamed of building a southern rival to Jefferson's gardens at Monticello, planted his experimental orchard, vineyard, arboretum, and show plantation just down the road. My own family has farmed the silts deposited along the banks of the Savannah River for 200 years.

There, they do what mulch is supposed to do, but you don't have to look at them. Cypress leaves make a delicate, lacy mulch of rusty orange. Bamboo leaves work, too—you can easily rake them up. But the ultimate mulch makers are the trees right overhead. In a tiny urban garden, I once chose all the small trees based on the composition of colors and texture of the fall leaf drop. I found a carpet in the house, took a picture, then walked through gardens looking for leaves colored to match that carpet. The idea was that the patio garden would be mulched in the same colors as the living room carpet.

The most important mulches, from a management standpoint, are our green mulches. We plant these heavily: peas, peanuts, or parsley to shade the soil and smother weeds. All of those plants are then cut, in place, sometimes chopped, and used as mulch. We simply cut the stems at the ground with a machete, drop the dry vines or stems, and chop them in place. They make winter mulch that helps maintain heat, moisture, and winter weed control. The microbes and insects break it all down by May, and we start over. To help reclaim new land, we use an aggressive, tall-growing mum called 'Miss Gloria's Thanksgiving Day'. We direct stick cuttings (meaning we take a 6-inch cutting of a plant's stem and stick it in the ground, where it will root and grow next spring) in the winter, let them smother everything all summer, then chop and leave the stems for winter mulch. What's more, as this variety flowers very late in November, it provides tons of pollen for bees and flies. Removing nothing, adding nothing, we let every living thing be a part of the complex system that we can't really begin to explain.

The work we do in our nursery also offers examples of ways that we can be gentler to the earth. We want to understand, to respect and honor the

LEFT Ferns are a traditional plant for haulm. Bracken fern is an aggressive groundcover that can be cut to provide lightweight mulch to regulate soil temperature and provide potassium. In this garden, it's contained by a brick wall.

RIGHT Cordgrass is an elegant ornamental grass that can be cut to use as winter mulch.

Though this garden might look like an old-style shrub garden at first glance, I added many multiuse plants, ones that fix nitrogen and many edibles including figs, pawpaw, pecan, and quince. Tom Hall and I designed and oversaw the renovation of this, the oldest house and garden in Columbia, South Carolina, to keep the spirit of the yard, a place that's had multiple uses for centuries.

life in soil, in the plants and insects—we want to share in the recognition that we are all part of one big system. In our nursery or in your garden, that understanding, that cycle, is perhaps the ultimate function of our plants—they are teaching tools. The techniques we've developed are perplexing to new gardeners, who are always asking why we do it. Professional horticulturists want to know if the interplanting just gets in the way and makes things more difficult. The first thing I always tell them is, "You can do it, too. The why and the how come later. Just try it in a little area."

Roger Swain, host of PBS's *The Victory Garden*, and I once had a discussion about the limitations of this sort of interplanting. He explained that home gardeners may plant some blueberries next to their roses, but when it comes down to protecting and seriously cultivating for harvest of berries, the gardener simply doesn't need to expend the resources. In other words, if you are not home when the flock of berry-eating birds comes through, you'll go to the store that afternoon to buy a quart of berries. That crop of berries isn't valuable enough to you to offset the cost of staying home to protect them.

I understand the practicality of that. There's no fun in having to go out on a dark, rainy night to pull up muddy beets while balancing a flashlight and wondering if you even have enough time to preheat the oven so you can deliver roasted beet salad, as promised, to the dinner table. But yet there's so much more to it all than that. Here we are, gardeners, volunteers, and high school principals able to reclaim a bit of dirt and make it do more than just look pretty. If you've ever worked the ground, even just a little bit, you know that satisfaction. And while your children may be bored when you garden with them now, that's okay—they'll remember. And they'll need to remember, because they'll eventually be the ones voting on agricultural policy and making our food decisions in the coming years. To develop a lifelong love, a deeper commitment, people need to be around people who use plants. Richard Hager, our cottage gardener and a lifelong professional educator, told me:

Fig preserves will forever remind me of climbing fig trees, itchy arms, and my mother, who makes them every year.

> Being around people who care for plants, especially plants that we eat, may not seem to influence a child, even as they grow up, even as an adult, for a long, long, long time. But it will. Give children a responsibility in the garden, and they'll get the spark. Anybody who was around it, one day will remember, will love to see things grow, to help things grow. Like all nine of us who lived in a little shack in North Carolina, they'll be a farmer at heart.

For my entire life, my mother has harvested figs and made fig preserves that we could use throughout the year. Her figs, roses, and azaleas grow together—it's a beautiful, coarse, textured backdrop to her flower garden. Figs require care just twice a year. In the spring, we pull out weed tree seedlings and vines and, in drought years, we water to stimulate fruits. In the fall, when we rake other areas, we dump the leaves under the fig trees. Otherwise, they need no pesticides, no fertilizer—nothing is added. The output is so much more valuable than the cost or the physical input. To be honest, I don't even like fig preserves all that much. But my soul has been fed forever by that connection, that memory, that fig tree. As Richard says, "Sure, I could buy corn and figs cheaper than I can grow them. Then, though, I wouldn't have all the extra to give away and the joy in giving!"

On our farm, we select plants that fill multiple needs, and we mix together all kinds of growing, living things. We stack. And we do so for all the reasons that cottage gardeners, everywhere, in all times, did it and still do it—it just makes sense. It's efficient, and it honors the earth's resources. Today's gardens don't have to be our *entire* pantry, medicine cabinet, hardware store, or art gallery, but they can contribute to all of these things. They can give us cleaner, deeper lives with more layers and more hope that in the future, the people who make decisions about what we eat, how we treat the world and each other, have some inspiration—that spark—from their own gardens, and the gardens of their youth.

Persimmon fruits are an added benefit to growing this elegant small tree.

Our Favorite Edibles

Andrew Keys, a writer, designer, and lifelong gardener, and the author of Why Grow That When You Can Grow This?, *says:*

Your garden's foundation is likely a combination of these botanical components selected for beauty and durability only. If you're ready to make a change, consider switching the for-your-eyes-only foundation out for some of these sexier, tastier substitutes. Remember to plant multiples for fruit crops, and get your soil tested to be sure it's safe for growing food.

Together, Andrew and I created some suggestions of our favorite edible plants that you can substitute for typical foundation plants.

Easy-Care Edible Fruit Shrubs, Trees, and Vines

SCIENTIFIC NAME	COMMON NAME	FAVORITE VARIETIES	NOTES
Acca sellowiana	pineapple guava		Edible flowers and fruit.
Actinidia	kiwifruit	'Ken's Red'	Small fruit on a large, shade-producing vine.
Asimina triloba	pawpaw	'Shenandoah'	Large tree with golden fall color.
Citrus ichangensis × C. reticulata	yuzu lemon	'Yuzu'	Cold hardy into the low teens. Used as zest for drinks and in cooking.
Diospyros	persimmon	'Maekawa Jiro'	Large shrub to 12 feet.
Malus	apple	'William's Pride'	Small tree with early summer apples. It has few pests, but is not self-fertile; you'll have to pair with another apple.
Mespilus germanica	medlar		Easily trained small tree.
Sambucus	elderberry	'Nova'	Great flowers for tea. Elegant leaves. Very aggressive.
Ziziphus jujuba	jujube	'Li', 'Abbeville'	Small tree with elegant, glossy leaves.

Building Fertile Soils

Encouraging a Healthy Web of Life

I GREW UP AROUND a bunch of farmers and gardeners who would often seek out the advice of the county extension agent. They would then listen, consider, and merge those technical instructions with the ways they'd always done things. They bought bags of smelly fertilizers and applied them at consistent rates according to chemical-based soil test results. They went to little meetings at the experiment station to learn about new peas. But then they'd spend half a day going into the woods, digging up worms and good, black, rotting leaf duff, hauling cow manure in buckets, or raking up some moldy hay from a truck spill off the side of the road to put around their cabbages and camellias. These were practical people—they mixed the old and the new to make their gardens better.

Sometimes, when they weren't trying to teach me about the joys of growing okra, they would take me climbing in oaks pruned low by sand and salt-filled ocean breezes, or exploring in swampy ponds filled with waterlilies, ferns, and pitcher plants. Besides making a gardener out of me, they left me with an enduring respect and love for my elders. Today, sitting with an old man on a tailgate and hearing tales of how he started growing vegetables and plowing with a mule, then flew planes over Germany, fell in love, worked twenty years printing the first phone books the world had ever seen, and now volunteers with the local extension service, I'm entranced. Most of the people I interview in this book could not have imagined that they would be where they are today, given their rural roots. These sparkling, can-do, thrifty, wrinkled-with-laughter children of several depressions and wars started out in a country that was stuck deep between a rock and hard place. That predicament had a lot to do with food, gardening, and soil fertility.

Practical, rural people make life even easier by establishing gardens near their fertilizer source. Here, old blush rose climbs a crepe myrtle with love-in-a-mist below and Brangus cows in the pasture beyond.

Where I'm from, the South, an entire generation and their children—actually anyone alive from 1840 until 1942—were touched by soil depletion. Decades of greedy farming practices, especially growing repeated crops of cotton, ruined the soil. Rich people, like cotton planters, lost everything. Some of them drank themselves to death, while some of them kept pushing on, growing just enough food for their families and farm workers. All the while, the poor hunkered down in shacks with rows of collards. Huge population migrations took people to other states. The cotton farmers on one side of my family lived in a crumbling plantation, selling off bits of land to stay afloat. The sharecropper side had to move in with relatives. Think about how that poverty of the 1920s affected your family. All across the country, soil fertility problems led to illness, a generation of smaller people, and hunger.

Synthetic fertilizer then came along and changed everything. For most of us now, it's hard to imagine a time when we couldn't go to the store and stock up on bags of those magic pellets. But back then they didn't exist. Instead, nitrates and guano came bagged from mines in Chile. Phosphates were mined and bagged in South Carolina. Yes, you spread them with spreaders, much like today's fertilizers. But complete synthetic, pelleted fertilizer came on kind of suddenly; the transition is literally referred to as the Green Revolution and credited with producing more food, fiber, and fodder

Modern construction methods often leave sites stripped of all topsoil. This is the original construction site of Riverbanks Botanical Garden in South Carolina, where we would soon apply many old-fashioned soil-building techniques.

on once ruined lands. But all the promise of these developments came with some serious drawbacks. And for many, these were only discovered in time, with advances in microbiology and hindsight. You'd think we'd have learned. But today, fertility issues affect us in *exactly* the same ways, though the effects can be masked and abstract for the casual observer—but we all feel them.

If you garden or grow any plants at all, you're closer to fertility issues— you know that we have to supplement the soil if we want to have healthy plants. And I believe it's time for a change in the ways we get nutrients to plants. Some of these are new ways, only recently understood by science. Some are old-school techniques used but later put away by our grandparents. We can fuse the old and the new to move forward. As gardeners, landscapers, and horticulturists with a deep connection to the dirt, we must lead that change. We must help others understand the impact that our actions are having on our soil. We can do better; we can fertilize our plants, camellias to cabbages, using natural cycles that get nutrients to plants and can lead us and our children to a soul-satisfying, healthier world of vegetables, flowers, and trees.

The Teachers

FRANK ATKINSON AND YVROSE VALDEZ

My two teachers here couldn't be more different, but they've both had an immense influence on me at very different times in my life. Mr. Frank hired me for my first job, working on his old-style farm that today might be called a polyculture (using multiple crops in the same space) farm. I met Yvrose

Valdez via social networks and friends of friends—thirty years later and many miles away from Mr. Frank's farm, she was using many of the same principles in her suburban garden.

Frank Atkinson, or Mr. Frank, as we knew him, was one of my first teachers. I can still see his face, furious and totally perplexed, when I, as a teenager, drove his tractor straight through the side of his barn. And I still think back to his satisfied expression when I learned to separate the thumbtack-sized clover seedlings from the weeds—in a field he'd let me seed all by myself.

Mr. Frank was a farmer, and a John Deere man. His farm was at the end of a sandy lane with two tire tracks and Bermuda grass in the middle. Sometimes, to give the kids a thrill, my daddy would drive us in the bed of his 1969 Dodge truck fast down that lane, bouncing off all the little humps. We loved it—that was often what passed for excitement for children from rural flatlands. Along the sides of the road, where truck tires pushed the sand up into long rolls, we made toad houses around our feet in the summer and picked from the blankets of fuchsia, red or pink phlox, and crimson clover that came up in spring. The road passed the Atkinson compound of three houses, trailed off into tracks through cow pastures, orchards, and barns. I don't know why, but the local Avon lady lived in one of the houses. Mr. Frank and his family were in the new house, and his "mudtha," as he'd call her, lived in the old house with the big front porch. Her garden was full of daffodils in the grass, spirea, thrift, iris, roses, camellias, and a low-limbed magnolia that was epic for climbing. Beyond, rolling hills of Bermuda hay,

A sandy lane leading to days of adventure and education on a farm like Mr. Frank's.

beef cattle, a you-pick scuppernong vineyard, and pecan trees made his a diverse farm with steady stream of income throughout all four seasons. He was a wiry, quiet man who never looked comfortable in either the John Deere jumpsuit or his Sunday suit. And for all the food that came off that farm, for all the flowers, for all that production, there was not, on the entire farm, a fertilizer storage room.

As a teenager, I was the farm boy. It felt then, and now, like an honor, a special position of trust and responsibility. Even today, when I see my predecessors, we talk about what a special job that was. Mr. Frank picked his boys carefully, and then he trained us and kept us around. He was a son of the depression; a gentle, reserved man whose idea of an off-color joke was anything including the word *tallywhacker*. All of us farm boys were sure he knew just about everything there was to know about engines, cows, plants, and barns. But he *didn't* know that sometimes we pulled the truck or tractor up to an electric fence and entertained ourselves by waiting for him to lean against it. Then he'd cuss—well, he'd say "dern"—and shake his tingling hands out while we tried to hide our amusement.

Despite our antics, Mr. Frank was a serious and reticent man. And I know that he and his generation tried to do the right things to repair the

BUILDING FERTILE SOILS

LEFT Crimson clover and oats build organic matter and nutrients in the soil at the University of Tennessee Botanical Gardens.

RIGHT In this new field for producing crinum lilies and purple pineapple lilies, we use quick-growing oats for their massive root system, which tills the soil and deeply incorporates organic matter.

land and leave us with something better. When someone showed up with that synthetic fertilizer, promising green again, it must have seemed like magic—Jack and the beanstalk stuff. People like him across the country must have imagined a world with no more mules pulling carts of manure, no more dust storms coating their houses, and, most importantly, no more undernourished, yellowing plants in failing fields or gardens—a green revolution, indeed.

Mr. Frank certainly did use synthetic fertilizer. I remember the instant jump in growth, too, after the nitrogen truck sprayed the Bermuda hay fields.

The Soil Catastrophe of the South

While most of the United States first felt the depression in 1929, it started nine years earlier in South Carolina with the crash of cotton. Many economic reasons combined to cause the crash, but one of the problems with cotton was that the plant takes a lot out of the soil. A common practice was to clear a new field and grow cotton for a few years until the soil was depleted, and then abandon that field. This created an abundance of empty fields that just sat around and slowly eroded into the rivers. This soil catastrophe began a few years before the Dust Bowl. Both of these events are among the largest human-made environmental disasters the world had ever known. And both had similar causes: poor land management, poor understanding of soil science, and, quite simply, greedy farming practices on a huge scale. While this happened in both Georgia and Virginia, it was most widespread in the Carolinas. This led to mass migration—the 1850 federal census showed that more than half of the population of Alabama had been born in South Carolina. By 1920, 8 million of 19 million acres of farmlands in the state were declared "Destroyed."

But he also used the lessons of soil conservation. And the reason there wasn't a fertilizer room on his farm probably had more to do with his thriftiness than any concern about safety. Why buy fertilizer, when you can just open a gate and let the cows clean up a field and manure it at the same time? Why buy fertilizer, when you can intercrop red clover under the pecan trees to enrich the soil? Why buy fertilizer, when beans inoculated with fungi spore produce more? He taught me to do all these things; he knew they were a part of the fertility cycle that conserved and created great dirt.

Yvrose Valdez, a wise, soulful gardener in Miami, has a stunning garden that reminds me of all that I've learned over the years from people like Mr. Frank. The road I took to meet her was a glaring and white-hot, five-lane city highway through Little Haiti, the poorest part of Miami. I had intentionally walked it for a few fascinating miles to get a feel for the neighborhood—but let's skip to the moment that the hot sidewalk was subsumed by a garden, a food forest. Everything suddenly became cooler, greener, and chirpier. Though I'd never been here, I knew I'd found what I was looking for. Even the air felt alive and humming with energy. In my head, I greeted and thanked the unmistakable life force of healthy plants. There was no need to ask, and no need to check the house number; this was my destination.

Yvrose, a friend of a friend of a friend, hugged me and showed me her garden. It's situated on a small city lot, and it's so intense and sophisticated that we took all afternoon for the tour. She talked at length about food issues, health, and how US farm policy affects nutrition and education in Haiti, her home country.

She's a beautiful woman, with silver-streaked hair and a strong, youthful body; a caring, worldly, and hip manifestation of the word *grandmother*. And this grandmother never does what most people would consider fertilizing; she doesn't even add much compost. She lets legumes do the work. In shrub beds, pigeon pea is interplanted with mangos and mulberry bushes. Among flowerbeds, perennial peanut spreads its yellow flowers prettily, making a tight little groundcover, all the while pulling nitrogen and feeding the other perennials.

These were *farming* techniques. Yet, here is Yvrose, a retired postal worker with a little backyard of flowers, using these old-school methods to feed her plants. I asked her why. "Because it works," she said. "Because

In the midst of otherwise typical landscapes, Yvrose's garden is an Eden, spilling around the neighborhood sidewalks. She uses permaculture principles to manage these beds, interplanting legumes to help add nutrients for bromeliads and fruit trees.

people who can't buy fertilizer can do this. Because this is how my grandmother, who taught me to garden a hundred years ago, in the most beautiful hilly farms of Haiti, would have done it." She's a teacher, as well. She volunteers her time helping others to start community gardens, and she knows and supports the local permaculture people.

Yvrose Valdez in her Miami Shores food forest garden.

Her garden is so beautiful, so attractive, and full of food that it never occurred to me that it could be considered controversial by anyone's standards. But it was. Yvrose says:

When I first did this bed, I had a little problem initially with the village inspector. It seemed wild. I told him, look, take me in front of the judge and I'll tell him I'm a retired person growing some food, and I keep it clean. No problem since. The other day he stopped to pick some lychee fruits! It was the first time he'd ever eaten a lychee!

That's Yvrose teaching again; evoking the spirit of her forebears and inspiring me to remember the lessons of my mentors like Mr. Frank.

Updates and Adaptations

On our farm, or on any farm, we are always taking things away, thus depleting the soil. Plants, vegetables, and flowers are made of dirt, sun, and water. So every time we harvest, we're selling a bit of our farm. We all do this. Every time we cut the grass, rake, eat a peach, or put on a cotton shirt, we've taken stuff out of the dirt. Plants are like little mining machines, pulling nutrients from the dirt. All the things we have, from medicine to smart phones, have some mineral from the dirt in them.

And since we take, we have to put back. For the past few decades, we've done that via bagged fertilizer produced in factories. Fertilizer did help save our soil, and it did feed the world. But, in hindsight, we know there were unanticipated repercussions. Almost no synthetic fertilizer puts back equally what we take out. So some micronutrients slowly deplete from the soil.

Plants That Add Nitrogen to the Soil

False indigo (*Baptisia* species). Spires of yellow, blue, or purple flowers in May. Steely blue leaves spring till frost. These long-lived perennials mix well with salvia or asparagus.

Japanese yew (*Podocarpus* species). Large, upright evergreen shrubs and trees.

'Little Volcano' bush clover (*Lespedeza liukiuensis* 'Little Volcano'). Dramatic fountain of purple flowers in fall, this 8-foot-tall shrub can be a backdrop for roses or perennials. It also makes a quick-growing, though slightly messy privacy hedge.

Powderpuff (*Mimosa strigillosa*). Dense summer groundcover with ridiculous little pink puffball flowers. Low enough to grow under shrubs or grasses, where it feeds them and blocks weeds.

Wax myrtle (*Myrica* species). Large evergreen shrubs that grow on poor soils.

White clover (*Trifolium repens*). Tiny white flowers and ground-hugging leaves can grow in your lawn, mixed into grass. In fact, this used to be commonly mixed into lawn seeds to help the tiny grasses grow.

You can also buy dried peas or peanuts from the grocery store—make sure they are organic and uncooked—and sow them in among your flowers.

We have to add more. The more we add, the more we damage microorganisms. We end up destroying those things like the fungal relationships that allow Yvrose's beans to pull nitrogen from the air. We stop the natural systems that change rocks, minerals, and even plant parts into soil nutrients.

Think about what lies under the dirt way, way down: rock. As that rock weathers, it becomes soil. But that change is more than just weathering. Naturally occurring fungi, bacteria, algae, and microscopic worms break down rock further, into nutrients the plants can use. Today, soil scientists understand their functions and fragility so much more clearly than they did just a few decades ago. In fact, soil microbiology is a whole new science. We know that factory-produced, salty fertilizer and pesticides kill these microorganisms and stop the conversion process. The cycles are all tied together, which explains why it is popularity referred to today as the soil food web.

Sensitive briar and four-wing bean can pull nitrogen from the air into the soil.

We look for ways to get those natural cycles rolling again. We want to help and encourage the web of life in the soil to provide fertility to our plants. One easy way to do this is to do exactly what Mr. Frank and Yvrose taught me to do: interplant plants that have the beautiful ability to provide nutrients to other plants. These plants, mostly from, but not limited to, the bean family, can pull nitrogen from the air, store it in their roots, and release it into the soil. In between our lily rows, purple peas, yellow flowering peanuts, and lima beans do just that. We overseed with clover and add flowering indigo to shrub borders. In new gardens that I make for clients, I seed in clover or peas. Though some people think of clover as a weed, it's an attractive plant and a great conversation starter. And since my day with Yvrose, I've included legumes in *every* garden and landscape I've made. From little white clover to huge Kentucky coffee bean trees, I seek plants that feed other plants.

But there is one caveat. For the most effective nitrogen fixation, you need to treat the roots of any new plant or seed with a fungal inoculant when you plant. This kind of helpful fungus—called mycorrhizae—really does all the work, and the spores of these tiny organisms may not be in your soil. The truth here is that this fungi, rather than the plant root itself, pulls nitrogen from the soil. Mycorrhizal fungi help us provide nutrients

Wingmen in the Dirt

We inoculate seeds with fungal spores to make sure young plants meet their beneficial part-
ners, but how do fragile plant roots make connections in the wild? Soil scientists now understand how symbiotic relationships initiate change almost daily. Imagine a brand new, tiny
soybean root emerging from a seed. If it can encounter the microscopic spore of beneficial
fungi, the plant can grow more vigorously. The new root produces chemical gaseous signals
that effectively call out to the fungus. But the chance of the two being in the same place at
the same time is slim, and producing those signals is both damaging and difficult for that little
seedling. However, that root *is* more likely to encounter bacteria. Fortunately, there are specialized, helpful bacteria that can detect fungal spores, and then communicate to the plants about
when to produce those chemical signals and which way their roots should grow in order to find
the fungal spores. In other words, there are wingmen bacteria!

The three—roots, bacteria, and fungi—then live together for the rest of their lives, each
providing something the other needs. Each different plant species in the world may have connections with different bacteria. There may even be specialized bacteria that help other bacteria set up profitable relationships—bacteria-brokering bacteria. In a single teaspoon of soil, in
fact, there may be up to 30,000 species of bacteria.

to plants. They come in many shapes and sizes—some even glow—but
mostly they are long, hair-like structures running through almost all healthy
soil. Normally, you can't see them, but every once in awhile, you'll see a mass
of fiber, fabric-like stuff in the dirt. I can see this covering on lots of my vegetable and plant roots at home, and the webby hairs can be yards long. And
yards of them can fit in just a single tablespoon of dirt. They intertwine,
penetrate, and help your plants' roots get nutrients.

Think of a plant's roots like your arms when you're trying to reach a
jar on the top shelf of the pantry that is just out of reach; your fingertips
touch it, but you can't quite pull it down. If you could somehow magically
extend your arm even just an inch, you'd have it. Well, for your plant roots,
that magic extension is often a fungus. They are like living pick-up grippers
for nutrients; they can extend the range of roots by up to a thousand times.
They're important tools for you to be aware of and cultivate for your plants.
For your flowers and shrubs, they're beneficial, but for your vegetables, fruits,
and grains, they're essential.

Besides physically extending the reach of plant roots, fungi also help
your plants reach nutrients in another way: they help with that natural

cycle, that weathering of the soil. Many nutrients are chemically combined in ways that prevent plants from reaching them. Plants simply do not have the ability to break down certain chemical bonds. Mycorrhizal fungi, however, are connected, living on the outside and even on the inside of plant roots, so they are a direct delivery system of these nutrients. Those little hairs get nutrients from wood, bones, and little rocks. They can even trap and digest tiny worm-like animals—these are all things that plant roots cannot do. A hunk of rotting wood in the soil may be full of

The white webbing—mycorrhizal fungi—on this crinum bulb is microbial life that helps the roots find nutrients.

nutrients, but plant roots cannot wrap around it and suck out what they need. Some nutrients need to be changed into other forms in order for plants to make use of them. Mycorrhizal fungi exude compounds that can break down those chemical bonds and then deliver those nutrients to plants. Fungi convert phosphorus, one of the most important nutrients for plants, from a form plants can't use, to a form they can. Fungi also have access to lots of micronutrients, including zinc, copper, iron, and calcium, which are in the soil but inaccessible to plants by themselves due to constraints of chemistry.

There are microscopic worms that eat the bacteria and fungi. And there is bigger stuff that eats them. Throughout their lives, they are excreting—through sex, birth, sickness, and death—always adding more nutrients to the soil. All the sticky stuff of their lives holds the soil together. If we all do

Farmers have long used inoculants to stimulate growth of beans. I use a tested bacterial and fungal inoculant, shown here on spinach seeds.

Conservation of Fungal Diversity Under the Jungles in Madagascar

Chris Birkinshaw explores plant diversity in Madagascar, from rosewood jungles and screw pine swamps to aloe-covered quartz mountains. Chris, an explorer, botanist, and swashbuckler type, is the hands-on director of the Madagascar Research and Conservation Program of the Missouri Botanical Garden, a world leader in conserving diversity. Protecting genetic diversity now extends *way* below the dirt. Chris and his team investigate the relationship between aboveground diversity of higher plants and belowground diversity of microbes. Botanists inventory the plants from ¼ acre plots, while the microbiologists measure diversity of microbes in the soil within the same plots.

Chris says:

> If we do find a correlation between diversity of plants and diversity of microbes, then this would suggest that a good way of conserving microbial diversity is to select areas with high plant diversity. The underlying idea behind all this is that humanity should try to conserve microbial diversity because, although a small number of species cause diseases, most are very useful in one way or another—including drug discovery.

The forests of Madagascar are in imminent peril. Chris Birkinshaw (left) spends days in the wild, working with local farmers to identify the most diverse forest. Diversity in plant life may indicate diversity in microbial life in the soil.

our part to encourage that underground life, all that sticky stuff can also be the building blocks of our gardens and lives.

Our crinum nursery is an old-style field nursery, where everything is grown in the ground, in rows. Tom and I do everything in our power to care for and encourage these natural cycles in the soil. And while we have seen a very slight slowing of our production, we take comfort in the idea that we are eating and living off a system that is better for both the earth and ourselves. And while we're not quite there yet, our ultimate goal goes beyond just giving up *synthetic* fertilizer—our aim is to not bring any fertilizers *at all* onto the farm from outside. But to do that, we'd need more animals, equipment, or labor. We still use a seaweed-based fertilizer twice each summer, which keeps things growing at a good pace and strikes a balance between the desire to be purely site sufficient and the reality of running a nursery where we need to keep our plants growing at a production pace.

Vegetables, herbs, and flowers are only the visible part of our crinum farm. These plants help build up the health of unseen microorganisms in the soil.

Another update we've made to help stimulate the soil food web is a modification of an old-school gardening technique. Since we don't have the animals or the labor to regularly cover the entire field in compost, we make compost tea, which works as an inoculant. It holds an incredible concentration of bacteria and fungi, which we use to keep our soil life super active. We use a 55-gallon brewer, which requires a 24-hour brewing period, during which air and liquid temperature are monitored. Compost teas are living liquids, and thus must be used fresh. You can make similar compost teas and inoculate your soil even if you live in an apartment, or garden on a terrace twelve stories up. Bokashi is a very cool process for making indoor tea developed in Japan—a place where every bit of space and landfill is valued.

This compost tea foam smells rich and sweet, indicating the presence of proteins and carbohydrates.

The only other thing we add to our garden is organically grown Bermuda hay to mulch the fields. That hay feeds fungi, bacteria, worms, and, ultimately, our plants. One of the first jobs Mr. Frank gave me was to bale Bermuda hay to feed his cows. He left me with that legacy. He also taught me to consider the needs of each specific plant and to inoculate seeds with

Rolling big, round bales of Bermuda hay around the crinum farm. Bermuda hay is also available in easier-to-manage square bales.

tiny, dehydrated organisms that helped them grow better. The lesson for me was this: there are things we can't see that are crucially important to things we can see. Now we unfurl that same hay onto our new-style, no-till farm as a way to feed our soil, our plants, our bodies, and to supply gardeners across the country with cool bulbs.

Our farm is our small experiment where we can share sustainable gardening practices with others. On a larger scale, leaders in landscaping, gardening, and sometimes even corporate farming demonstrate that we can change our fertility systems. We can grow lots of healthy, productive plants that provide food, medicine, construction materials, and pleasure by under-standing and manipulating natural systems. This doesn't mean going back to mules and wagons, or to shiploads of bird poop sailing around the world. Rather, it's building on how the farmers and gardeners of just a few decades ago worked; by picking up the good bits and moving forward. With real inquiry, and whole science, we can provide nutrients to our plants through old methods and new techniques. The world of plants above and below the soil remains larger than science has yet to fully comprehend. As gardeners, we must embrace that mystery, that spark of life; we must honor it in every-thing we do outside.

Bokashi Composting

Despite what you might have heard about composting, you can indeed do it right in your kitchen. And you *can* compost food scraps, including cheese, fish, and cooked stuff, lawn trimmings, and pet waste—anything once living. A great way to do it is through bokashi composting, which employs specialized beneficial microbes to compost in an airtight container.

Developed in Japan in the 1980s, the bokashi composting method basically pickles your kitchen waste. You put everything in an airtight bucket—be sure to use a bucket with a drainage spout near the bottom—and add a mix of microbes and bran meal, which is available on the web or in specialty stores. Add scraps for a few weeks and then sprinkle on the meal. Then, from the spout, you can pour a nutrient-rich cup of compost tea. Mix that with water and use it to fertilize your plants. Every few weeks, you dump the compacted compost materials from the bucket and start again. I bury mine in the yard. A friend who lives in a 12th-floor condo tried this, too. At first he liked it; his patio citrus were definitely improving. He'd take his monthly bundle of compost downstairs in the elevator and bury it in a bed near the community pool. But he later admitted that the process was cumbersome and he wasn't fond of the smell when emptying the bucket. But you could simply throw the compost away, or if you live in a dense city, there may be a bokashi service that will come and pick up your compost and leave you with a clean bucket. Indeed, the process can be slightly unpleasant, thanks to the appearance and smell of what comes out of the bucket, but it's great stuff, and worms will flock to it. After a few weeks, it will turn into beautiful black dirt.

In my household in the city, we always have lots of fresh produce waste and a small yard. The bokashi method is perfect for us. And if you're concerned about the smell, don't be; neither the bucket or the tea smell very much at all when they're sitting on the kitchen counter (though you might want to avoid sticking your nose into the bucket). I never have any issues with flies or other insects. After using the tea for a few weeks on your potted plants, you'll see a marked difference in foliage color.

A glass of tea from my bokashi compost bucket can be mixed with 5 gallons of water and used as a fertilizer. Next to it is the bran that you would sprinkle into the bucket every few weeks.

Stop the Tilling Cycle

Harnessing the Natural Powers of Worms and Mushrooms

CIVILIZATIONS AND GENERATIONS—and lots of tomatoes—can fail when we neglect our soil. Geomorphologists, who study natural formation of land, say it takes about 7000 years for natural processes to build a single inch of topsoil. Typical, entrenched gardening, landscaping, construction, and farming styles, especially tilling, can destroy soil in many ways—and quickly. Even in a small vegetable garden or an intensively planted annual area, tilling the ground chops up and kills microorganisms, and stops the natural cycling of nutrients. For initial preparations, tilling and adding loads of compost is an expensive, cumbersome way to build soil.

Our environment can't afford the soil loss, the dirty streams and rivers. And in a tiny garden or a giant farm, we can't afford the loss of nutrients. Tilling contributes to infertile soils by disrupting natural nutrient cycles. And when we then try to remedy the problem by adding fertilizers, they too

This one-year-old garden, made on hard, rocky clay, has never been tilled.

are environmentally expensive, and synthetic fertilizers often cannot provide the same types of nutrients as the natural cycles. For flower gardening, this may seem like something of a philosophical distinction, but in vegetable gardening, it's critical to our bodies, society, and economy.

Unfortunately, people just love to till. Take my mother, for example. She's heard my preaching many times, but remains addicted to the satisfying purr of the rototiller and the rewarding view of a cleanly tilled bed. And we all know a tiller guy, too, with tiller toys in every shape and size. Want a new flowerbed at the mailbox? Weeds coming up around the tomatoes? Creating a whole new landscape around your house, store, or fire station? Going back to your roots and becoming a farmer? Tilling became a default starting point for any of these processes. There is some logic to this, as tilling introduces lots of oxygen to the soil, stimulating some soil life and making things easy for roots. But it's a short-term and addictive process; longer-term cycles are broken. I also see it as a psychological dilemma; tilling is satisfying, easy, and rewarding. It just *feels* like the best place to start.

It seems like tilling became an easy answer only after oil became less expensive, though this wasn't exactly the case. It did become really easy at that point, but with draft animals and plowing, we've managed to till for centuries. We've ruined soil all over the world. One of the main drivers of the expansion of the Roman Empire was the need for new dirt. They literally wore out their soils, and then moved on. At the end of the nineteenth century, we even had tractors powered by steam engines. But it was gasoline that made tilling an easy, weekly task. And that allows us to weaken soil more quickly than ever before.

Frequent tilling is too commonly practiced in gardening; it overaerates soil, eventually killing off the soil food web with which plants have symbiotic relationships. Here, Tom Hall does the dirty work of keeping my mother's small tiller operating.

Today, more than ever before, we understand exactly why overtilling reduces the soil's water- and nutrient-holding capacity. We understand that overtilling results in serious, though often unseen, erosion problems. Worms and microscopic life, especially fungi, are literally chopped to pieces by frequent tilling. Remember that it's their bodies and secretions that hold your soil together. They can't repair themselves before the next tiller chops them up again, and as a result, we're left with soil that does little more than hold plants in place. We now understand how to landscape, grow vegetables and flowers, and feed the world without ever breaking out the tiller. The more you know about what it takes for healthy soil, the more no-till gardening just makes sense.

But while most farmers have by now embraced no-till growing, gardeners and landscapers have been slower to give up the tines. In my gardening practice, tillers, plows, and double digging went the way of the mule. Engines frustrate me; I'd rather be barefoot and quiet. I'd rather smell the dirt and be one with it than fight with cords and fumes. Tillers annoy me and scare the dogs. In all sorts of gardening, farming, and landscaping projects, I've learned to work without them. I've learned to care for and harness the power of the huge team of microbes, mushrooms, and worms that build soil and make our gardens beautiful.

If you're starting a new garden or even just planting a few containers, you'll save yourself both time and money by avoiding the tiller as you get started. And if you have an existing garden, you can still change your ways.

Home gardeners and large landscapers and farmers alike are converting to no-till gardening. In all my experience as a professional gardener and nursery manager and as a home gardener, no-till works in every situation.

The Teachers
LINDA PROFFITT AND TRADD COTTER

In a natural system, mushrooms and worms are amazingly successful when it comes to adding fertility to the soil—beautiful, slimy creatures that work together to feed our plant roots. Old country gardeners often use a practice that I always thought might be a kind of witchcraft. When planting a new tree, say a dogwood, they'd go into the woods and get a bucket of duff from around the same sort of tree, which they would then spread around the new tree in the yard. My teachers here are well aware of how this process works, how it actually innoculates the new tree, which had been grown in sterile nursery potting soil, with all sorts of spores and microorganisms that help it thrive. Linda Proffitt is a worm expert and Tradd Cotter specializes in mushrooms, and while both are practically youngsters, they're masters at using old tricks in new, mindboggling ways to build soil.

Linda Proffitt is constantly telling you what she's thinking, dreaming, and scheming, all while she shows you some wrinkled plans she's drawn up,

Linda Proffitt gets donations of woodchips from tree crews that she and her worms turn into acres of rich dirt.

talking with great enthusiasm about changing the world; you might wonder if she's a visionary or just a romantic. A year later, she's done it all; her dream of a giant, urban worm composting site and non-profit community center serving thousands is in full flower and all over the news. And the base of her success is a rich field of topsoil that she created over an asphalt parking lot.

Linda lives and gardens in Indianapolis, Indiana, and she and her beagle, Pearl, took me on a tour and showed me parts of that city that were really unsettling. She took me to once sturdy, elegant neighborhoods, now almost entirely boarded up and abandoned. She showed me around her own neighborhood, where even today, arsonists target homes that people

Making Dirt in a Small Space

When visiting bucolic Bowling Green, Ohio, a horticulturist friend said to me, "I want you to meet someone who's gardening in the worst of conditions." We drove north for a bit, into a place I'd have to call a slum outside of Toledo. In an urban fortress, surrounded by abandoned houses, sitting on rock-hard, certainly toxic soils, we met Paul, who grows his own food in 5-gallon buckets. At first glance, I thought he'd somehow buried the buckets in the ground, but that would have been impossible. His backyard dirt was too hard and compacted to dig into, and he couldn't afford a tiller. Rather, he had taken the buckets, cut the bottoms out, and set them on the hard dirt. He added compost from his kitchen waste, some greens from the yard, some cardboard, and then planted and watered with collected rainwater from his roof. He's not just gardening in containers, though; he's using the circular form of the buckets to make contained environments conducive to healthy roots, worms, bacteria, and other microorganisms. Over time, those roots and microorganisms move down, creating deeper layers of great dirt. A few years later, he has about forty beautiful little circular gardens, brimming with vegetables and protected from the dogs that regularly run through his yard—all this while the healthy topsoil just gets deeper and deeper.

Paul uses the cut off tops of buckets to create pockets of topsoil in compacted, urban soil that he otherwise cannot garden.

are trying to rebuild and bring back to life. It's both frustrating and sad to think about why so many people would abandon and leave these once warm neighborhoods. But in between the hopelessness, on empty lots and corners, Linda pointed out vegetable gardens, soil that she'd reclaimed by encouraging people, by helping them find resources and hope. In her red truck, Pearl in tow, we drive by gardens peeking out from behind parks, churches, and schools.

Eventually, we stop at a giant parking lot surrounded by a few run-down, white warehouses. This is where she pulled out wrinkled plans for the headquarters and soil-building facility that she planned to call Peaceful Grounds. It's a nonprofit based on the belief that unless people's most basic needs are met first, we will not be able to address any of our other problems. Then, the place seemed desolate and hopeless, but in the years since, Linda has already built up two feet of topsoil. On that former parking lot, she is planting vegetable gardens and mushrooms. More importantly, she's shar-ing the secrets of how she got there, of that new dirt, with teenagers from all over the world. Linda's ultimate message is that you can do it, too—in a yard or in a square foot on your balcony. I ask Linda if I can refer to her as a vermiculturist. She rolls her eyes and corrects me:

> Just say I'm an old hippy. In the 1968 Earth Day Parade, I won
> an award for decorating my bike. That's the kind of thing that
> changes your life. My grandfather, father, and uncle were preach-
> ers and gardeners. My uncle grew a garden and took food to
> shut-ins. I started out making gardens on abandoned lots to feed
> the homeless, and now, I make topsoil that will make more food.
> And, shoveling it around with kids lets us talk about how much
> soil we lose every year, and how much we all need to be resilient,
> to keep up some knowledge of food growing in case there comes
> a food emergency.

Besides being a garden, Peaceful Grounds is actually a massive topsoil production operation—and it costs next to nothing. Linda works with local tree companies who deliver and dump waste woodchips, covering the parking lot. She then adds compost inoculated with worms and worm eggs on top. Then she finds people to help her turn it, speeding up the soil-building process. Through a national network of people who share and admire Linda's vision, students come to Peaceful Grounds to work,

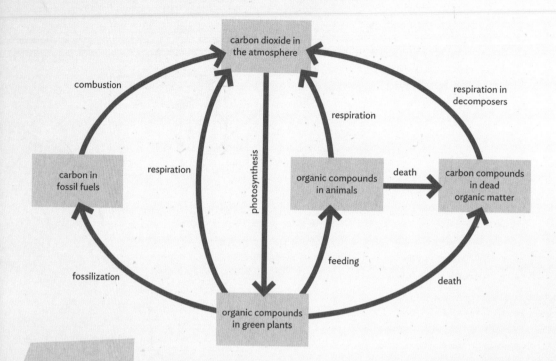

carbon dioxide in the atmosphere

combustion

respiration in decomposers

respiration

carbon in fossil fuels

respiration

photosynthesis

organic compounds in animals

death

carbon compounds in dead organic matter

fossilization

feeding

death

organic compounds in green plants

Virtually all life depends on the carbon cycle. In gardens, we can mimic and enhance the cycle by adding extra carbon from woodchips or newspapers, which soil organisms can turn into fertile soil.

learn, contribute, and turn all those woodchips by hand. In the first year alone, she had 2092 volunteer students from around the world who've donated 18,000 hours of their time. She worries about how we no longer pass down our gardening knowledge from generation to generation—but she's doing her part. And these volunteers do other community service while in Indianapolis. They garden, paint, haul dirt, and offer their services to churches, community groups, and others who are working to restore the city's warmth. At Peaceful Grounds, with little more than worms and woodchips, they've helped to achieve a dream. Linda gives this topsoil to people in the community who want to grow their own vegetables at home or in urban community gardens.

Linda's techniques translate easily to vegetable and flower gardens, landscapes, and small yards. A square yard of woodchips, with a little compost and worms, is all you'll need. Well, that and some sun, water, and air, and you'll need to turn the stuff a few times—everything that converts woodchips to dirt is alive, and they will need a little help to stay that way. As Linda puts it, "This is biomimicry, we're doing what nature does, speeding it up and using it."

In any new landscape, on your own driveway, or even in a huge container, you can use Linda's techniques, too. And not just with woodchips—you can stimulate worms with any cheap carbon source. Carbon is one of the four building blocks of life; it connects things. Nitrogen, hydrogen, and oxygen all love to bond with it, making up the structures of many different cells. Wood chips, bags of raked leaves, cardboard boxes, and piles of old magazines—these are all great sources of carbon, and we are constantly throwing them away. We are wasting the basis of great topsoil.

The cycling of carbon is absolutely critical to all life, and the carbon cycle can be manipulated to provide nutrients to plants. Plants absorb carbon from the air during photosynthesis. That carbon becomes, through photosynthesis and respiration, leaves, trunks, roots, and flowers. When any part of the plant falls to the ground, or when plants die, microorganisms in the soil go to work deconstructing the plant material. Some nutrients go into soil, and some go into fungi bodies and become rich organic matter and humus. In the initial phases of this decomposition process, organic matter is easily oxidized and carbon is released back into the air. The whole process continues, in a cycle.

The Power in a Termite's Stomach

Animals have flexible membranes around all of their cells, but plants have more rigid walls. These walls are layers of carbohydrates. Around them is a fluid called lignin. Lignin is what makes vegetables crunch; it's what we seek to eat for fiber, and it gives trees the ability to grow tall. Basically, the rigidity, the fibers of these two things make up wood as we know it. For simplicity here, I'm going to use the term *cellulose* to include both the walls and the lignin. Cellulose is energy, but it's locked up and unusable to plants and animals as food. Termites are an exception to this. Stanford scientist and former US secretary of energy in the Obama administration Stephen Chu once said that if only we knew how to use the bacteria that lives in a termite's stomach and helps them eat cellulose, we could solve our dependence on oil and address some hunger issues. Breaking that cellulose down into organic matter is a difficult chemical process that plants cannot achieve. Gardeners want it to happen fast, but we don't want termites around the garden, deck, or arbor. Fortunately, mushrooms are also an exception and can process cellulose rapidly. This is how mushrooms decompose woodchips, leaving us with a sort of raw topsoil. Then worms eat that, further breaking down nutrients into forms that plants can use.

When we interrupt the cycle, the worms, bugs, and tilth quickly go away. The soil then becomes compacted, and then we break out the tiller or the plow, which is part of its own vicious cycle that ultimately leaves us with lifeless soil. In a small garden, lifeless soil is frustrating to work with, and in a landscape, plants will look weak and malnourished. In large agricultural fields, rain and wind carry the dirt away with stunning, but often imperceptible, speed. Linda says, "The lesson I teach, while we're turning chips and worm colonies with pitchforks in our little garden, is that Indiana loses a trainload of topsoil every minute." It's a message that we have to keep passing down; we simply can't afford to lose that topsoil in any situation. Through her work, Linda is equipping our youth with the knowledge that former generations carried with them.

Besides plants and worms (and ourselves), fungi are also a crucial part of the soil building cycle—saprophytic fungi (think the mushrooms that we eat), to be precise. Tradd Cotter is a sapro-entrepreneur whose company is Mushroom Mountain. Down a wooded lane in the Blue Ridge Mountain foothills, Tradd and his wife, Olga, might at first appear to be just a couple

Take Back the Dirt

Remediation is the process of reclaiming, cleaning up, or creating new topsoil after the construction of a new suburban home, or even after something as catastrophic as a chemical spill. With the right selection of plants (phytoremediation) and fungi (mycoremediation), we can use the techniques of remediation at home to build up topsoil in our gardens and landscapes. In the 1930s, a key technique was to plant certain trees and groundcovers.

In the South, that meant pine trees, kudzu, and sea oats. Kudzu's massive root system held our abused soils in place. Its ability to fix nitrogen helped rejuvenate nutrition in the soils. While its invasiveness has become a problem, I tend to think we're hard on it. We brought it in, we enjoyed the benefits, and we still depend on the soil it stabilizes.

In the Midwest, to fight the dustbowl, shelterbelts of Osage orange trees were planted for miles to catch windblown soil. Sometimes these plantings stay put for decades; other times, they are cut and replaced after their work is done. Other plants can literally clean the soil of toxins. Ragweed, for example, can be used to pull lead from the soil, though in order to detoxify a site, a ragweed crop may have to be grown and harvested (and removed) for years. Fungi can serve the same purpose. Remediation can indeed work on a large scale, but it can be slow, and therefore difficult to write it into a government contracts, which tend to require defined timelines.

Mushrooms can be grown in and around plants to help turn carbon sources such as woodchips into nutrients that plants can use.

of young hippies experimenting in their own wooded garden. And they are, of course, but they're also leaders of this new movement of harvesting the power of mushrooms for food, medicine, remediation, and creating topsoil. He finds, grows, and even "trains" mushrooms for soil remediation. For example, for a soil that's suffered an herbicide spill, Tradd might train a mushroom to clean it up by exposing hundreds of mushrooms to the chemical toxins and then waiting to see which ones die and which one literally eats the toxin. That mushroom can then be used to clean the contaminated soil, making it safe and inviting for worms and other soil builders. Tradd says: "Feed them a little cellulose, and they are like the guest who brings wine to dinner. They make it easier, quicker, and more fun for all the living things needed to make topsoil. Once they colonize, they exude enzymes that taste sweet to worms."

I know that some people are still turned off by the sight of mushrooms in their garden; this process isn't yet for everyone. But, inspired by Tradd, I've been experimenting with ways to help people see their charms and beauty. I've been trying to pair them with perennials—color echoes, textural contrast, and other garden design tactics using mushrooms. It's all really new, but this year, I did love the coffee-colored skin of king stropharia mushrooms against the blue-gray leaves of elephant garlic. And I

Mushrooms: Some You See, Some You Don't

Estimates suggest there are 5 million different species of fungi in the world. All mushrooms that you see are fungi, but not all fungi produce aboveground mushrooms. Scientists have classified them into different groups. Plants depend on many of these groups, but two in particular are very important to gardeners.

Saprophytic fungi are the mushrooms we most commonly eat. They grow in organic, carbon-filled material like old logs, piles of leaves, hay, or sawdust. What we see aboveground are the reproductive structures. One single mushroom body can cover hundreds of square feet underground. As we've already discussed, these are terrific soil builders.

Mycorrhizal fungi form a partnership with plant roots, and we rarely see them. They wrap their hyphae into and around rootlets. Even under a microscope, it's tough to tell the root from mycorrhizal fungus. Roots supply the fungus with moisture and carbohydrates. The fungus becomes an extension of the roots, supplying roots with minerals and water that are physically or chemically beyond the plant's reach. In gardening, we use mycorrhizal fungi to help plants grow better.

like the contrast in form of the almost perfectly circular disc of the mushrooms against the strappy leaves of our crinum lilies. Given that the mushrooms last only a few days, the beauty is fleeting, but that makes it all the more special.

Of course, the real reason I'm growing mushrooms with flowers is to build soil fertility for the perennials. In one area of our crinum lily fields, the roots from nearby pecan trees encroach, leaving the soil dry and depleted of nutrients. Our crinum lilies grow more slowly there, so we're testing a huge mushroom bed in that spot as a way to keep the soil fertile. We ordered king stopharia mushroom spores from Tradd, which arrived in the mail, mixed into a block of moist sawdust. Following the instructions, we spread that over the ground, then layered woodchips, more spores, and newspaper and soaked it all down. We watered every day that it didn't rain. Within a week, white hairs—called hyphae—had spread into the paper and wood chips. Watching this process was an amazing, sublime experience, like watching leaves on a new plant unfurl. In just under three months, the stunning, wine- and coffee-colored mushrooms—the reproductive structures—erupted. We used them in our omelets.

What more could you want than something that lives in your garden, making rich soil from paper, while being both beautiful and tasty.

You've probably seen the work of colonizing mushroom hyphae already. It looks like a mass of cotton candy or tiny strands of a spider web. You'll see it under mulch, woodchips, straw—even pine straw—or in an old leaf pile. Mostly, though, it's underground and unseen. Eventually, the fungus reproduces, sending up mushrooms. Then it dies, too, becoming a part of the recycling of nutrients.

Often in new gardens or new suburbs, you'll see an outbreak or bloom of mushrooms. It is especially noticeable when they pop up in lawns, forcing most homeowners to try and figure out how to get rid of them. But we should learn to love and embrace them, as they are the beginning of soil building. The roots of trees that used to be there and the buried construction trash have become food for the growing body of fungi. When it reaches a certain mass, it sends up mushrooms. This is a good thing. This is the first sign that your soil is coming back to life.

For millennia, fungi have been doing exactly what we need them to do: eating dead wood and turning it into living topsoil. But as much of our soil has been ruined, and the carbon sources have been removed, the population of saprophytic fungi dwindled. The mushrooms can't create topsoil

by themselves. You can't just scatter mushroom spores onto compacted soil to get dirt—they need to eat. And to eat, they have to have some helpers. Therefore, you need all sorts of other things like worms to build soil. As Tradd says, mushrooms get the party started and attract all the worms and other important soil builders. But if the saphos do the initial colonization, their sweet enzymes attract other soil workers.

Linda and Tradd are shining examples of how to take basic knowledge and combine it with the latest in scientific research to move forward and make better soil. Through their work, they're helping to create new businesses and new lives for people by accelerating the creation of topsoil. We can do this, too; we can stop the tilling and support landscapers, farmers, and new businesses that strive to garden with the web of life.

Updates and Adaptations

On the edge of our little crinum farm, you can see and step down a startling 8-inch drop-off. It's the line that divides our farm from our neighbor's. That's 8 *inches* of topsoil—soil that took the earth millennia to build—gone, lost to wind and water erosion. In the face of this glaring evidence, our neighbor has recently made the conversion to no-till agriculture. And for the first time in my life, I don't see clouds of beautiful red dust drifting from his fields over to mine.

Our farm hasn't been tilled in more than thirty years. That was when my father turned it into Bermuda pasture. Not because he had a vision of till-free farming, but because he had limited resources. Bermuda hay production doesn't take lots of tractors, giant tillers, or other equipment. As Tom and I have needed more space for our field nursery, however, we've had to convert that Bermuda grass pasture into growing fields. Bermuda's tenacious roots go deep; they propagate from even the tiniest living bits of roots, even parts covered by inches of soil. It even withstands the toughest of synthetic chemicals. But it does make a great mulch; and even if there are seeds, we mulch so heavily that they get smothered.

For this meadow planting, all grasses were presoaked in a mycorrhizal inoculant.

Complete conversion from pasture to usable garden soil takes about a year, but it takes even longer to build a rich, friable topsoil. The techniques that we use can be used on any existing lawn. In late winter, we cut the grass as close to the ground as possible, and then lay compost and woodchips over it. Any newspaper, magazines, or cardboard that we can get our hands on gets put down, too. We don't even shred it. We top everything with dry Bermuda hay, which is our main carbon source. For the most part, that's it—

Plant Grasses in Spring and They'll Jump in Summer

Contrary to the general garden rule that says fall planting is best, grasses and a few other plants should be planted in the spring. All plants are divided into two groups based on how they utilize carbon during photosynthesis. This physiological difference in how they work is invisible to us but significant to the success of our gardens. One group includes most common perennials, trees, and annuals. They grow best in moderate temperatures. The other group includes grasses and other plants that dominate warm zones, including tropical grassland savannahs. They grow more efficiently in heat and drought extremes. In fact, they need heat and warm soil to grow rapidly; in cool, wet weather, they slow down. Most ornamental grasses that thrive in warmer climates are part of the second group (think corn and sugar cane). Those should be planted in the springtime—as the soil and air heat up, they'll start growing rapidly. A tiny plug of warm-season grass can become a 3-foot clump by midsummer.

The Other Side of Worms

When we scattered worm-filled compost, we were also scattering worm eggs, inoculating the soil further. If you don't have access to this, you can simply buy worms (along with their eggs) as bait at fishing stores. A plastic cup of worms stored in a cooler is a modest life for these underground pioneers and missionaries—civilization builders. They're little colonists who quickly populate entire new worlds, and turn inhospitable lands into places full of opportunity for other life forms, especially plants.

I've been designing gardens since the 1980s. Each garden or landscape depended on maintaining or building great, healthy dirt. I've built topsoil on rocky red clay at the Riverbanks Botanical Garden. I've planted a pleasure garden in Managua, Nicaragua, on pulverized, compacted volcanic stuff that looked like crumbled asphalt. I've worked on ashtray-like sand dunes in Florida, on limestone bluffs, and on my sandhill garden. I've worked on mucky clay near Puddin' Swamp, South Carolina (the name says it all), and in my own field nursery. In each, and in many more, I invited earthworms.

We should, however, recognize that worms don't benefit every situation. Keep in mind that many earthworms were introduced to America by Europeans, and new species still arrive. They adapt to urbanized soils better than insects, and they change soil structure—which isn't always a good thing. They changed our forest and may threaten wildlife. As much as we need their tilling in some places, many forest and landscape types, such as arid gardens, do not need worms.

After a construction project on our barn, the soil was compacted and lifeless. Here, volunteers add worms to exposed red clay, which the worms will soon turn into topsoil.

that sits all summer. If green grass peeks through, we cover it. Sometimes, we make little planting holes and seed in gourds or running beans. This gets roots growing in the soil, helps shade out weeds, and gives us a place to plant fun things like giant gourds. Normally, after doing this for a year, we can start planting perennial crinums, which are strong, tough bulbs with lots of foliage that help shade the ground.

I take these soil-building lessons off the farm, too. In my garden design work, no-till gardening solved the problems of some of the most challenging, hard-packed, rocky soil I've ever dealt with. Just outside the gates of the famed Augusta National Golf Club in a traditional neighborhood of brick walks, azaleas, and zoysia lawns kept perfectly edged, I was asked to design a wild meadow—complete with sheep.

The soil was no more than a quarter inch of red clay over rock. Other landscapers had started and abandoned the job. Half of an irrigation system was installed, but in some places, pipe was literally laid on top of the ground. There were lots of challenges around this newly built home, but the biggest was the lack of soil. You simply couldn't dig a hole for a landscape plant. A typical till approach would have been impossible.

Warm-season grasses in meadow gardens look sparse in spring, so I overseeded this meadow with spring ephemerals like these Chinese toadflax (*Linaria maroccana*). As the grasses grow in with coming heat, these delicate flowers dry up and fade away.

Fortunately, no-till gardening worked beautifully. With mattocks and short-handle spades, we dug holes 2 inches deep—and sometimes we couldn't even get *that* deep. In those divots, we planted thousands of plugs of ornamental grasses. Their 2-inch root balls sometimes had to be split and formed around rocks. Each grass was soaked in a vat of water and soluble mycorrhizal inoculant. Once planted and watered in, we tied up the grasses and topped it all with a very light leaf compost that was full of earthworms. A few months later, we followed up with another top dressing of compost made of vegetable matter and woodchips, along with a second inoculation of mycorrhizal fungi.

Essentially, we replicated the natural process of soil building. Now, the roots of the grasses, the underground fungi and bacteria, the mushrooms, and the bugs stay busy breaking down the heavy clay, opening it up and storing organic matter and nutrients in the soil. In just six months, our work grew into an elegant, flowing garden. A friend described the stone house on the property as floating on a pillow of grasses. Three different evergreens—bullrush, cordgrass, and lyme grass—make up the mass of the planting. Interspersed, winter-dormant sweet grass, love grass, and a few

In this meadow, we included elements to help "frame" the wildness, to let people know that this look is intentional. Formal stone walks, sheered hedges, stiff, evergreen plants, and bits of mown lawn accomplished this.

STOP THE TILLING CYCLE

others add multiseasonal interest. Gray leaves repeat throughout. The other plants, including a few shrubs, are all edibles: pomegranate, olives, pineapple guava, and the camellia that is used to make camellia oil. The walkway is interplanted with creeping mint, oregano, and blue flowering mazus. We've even planted annual salvia with tickseed mixed in. But overall, the look is of grassland—a wild meadow, manicured around the edges.

Why do it this way? A landscaper stopped by to look around and give unsolicited advice: "You should have just brought in 18 inches of compost and topsoil over the entire site.... They can afford it." Maybe, but I can't, and in the long run, none of us can. The way I look at it, I saved everyone a lot of money—and not just in the cost of soil or the equipment required to scarify the clay and to spread and till the soil. No-till saves all that, but it also saves the cost of fuel to do all that. And in this case, it saved a ton in the cost of chemical weed suppressant.

Tilling often creates weed problems by exposing to light weed seeds that are deep in the ground and dormant. All soils have a seed bank. Tilling opens the door to that bank. Bringing in and adding topsoil often does the same thing. Where do you think that topsoil comes from? It comes from places where someone is digging a basement or clearing land to build tract homes, and it often brings along its own bank of weed seeds.

We also saved the cost of repairing erosion issues on this sloping lot. There simply was no way to completely integrate a new layer of soil with the existing soil. The new topsoil would have floated on top, and our heavy summer rains would have gradually washed it down to the bottom of the slope, leaving little canyons, damage that would have to be fixed after each and every rain with costly, repetitive, frustrating hours of shoveling.

Perhaps unsurprisingly, in this tidy and conventionally landscaped neighborhood, our work with conservation and soil building has not pleased all of the neighbors. The meadow stands in solitary and stark contrast to all the golf course–like gardens that surround it. To my eyes, the contrast between the meadow and the wooded neighborhood is beautiful, much like a glade growing in the forest, where a ray of light snuck in. Here, gray-leafed grasses catch sun and wind. In midsummer, gray-white flower heads of love grass mix with a few little sparkles of red salvia and yellow cosmos. Later in the fall, sweet grass makes

I call this sort of walk from the meadow a goat-cart path, as it reminds me of places without cars, where animals pull carts on little tracks. This minimal hardscape allows for rainwater penetration and makes a place to grow mat-forming plants.

a mauve mist. Into that mist, on a rainy fall day, we sow the seeds of blue larkspur, white clover, and pastel toadflax. These germinate and grow under the grass and erupt into a spring show, just a little more organized than a natural meadow.

This kind of soil building can be a tough sell. Asking a homeowner to wait a year while some fungi bodies grow underground is quite a challenge. Writing contracts, for clients or a government agency, that specify mushrooms, massive seed-in projects, and slow soil building is very difficult—but not impossible. The typical ways we attempt to change soils, by adding loads of stuff and tilling it in, when repeated, are counterproductive. My daddy used to say, "That kind of thing is make-work," by which he meant, a never-ending cycle of creating a problem, then offering the solution. No-till vegetable gardening, landscaping, and farming works. Ironically, no-till farming has been commandeered by the chemical industry and has the reputation of being pesticide intensive. It doesn't have to be. If you learn to feed the living organisms that make up your soil, from worms to bacteria to fungi, you'll gain access to a joyful, hidden world and find a quiet, slow, and soulful way to love your dirt.

Partnering with Mushrooms in Your Garden

To have mushrooms, forget the shitake log or a wet bale of hay sitting around waiting to be watered. Think about it—mushrooms grow in the dirt everywhere. They can do the same in your garden. They can create spectacular colors, shapes, and forms that you just can't get from plants. You can mix them right in with perennials and shrubs. As plantings mature in spring and early summer, shade is provided for the fruiting bodies. But some mushrooms can be grown in full sun, too.

Helen Yoest, author of Gardening with Confidence: 50 Ways to Add Style for Personal Creativity, *gives solid advice to gardeners based on her hands-on experience trying and testing in her own North Carolina garden. Helen tells the story of her interest in mushrooms:*

In the fall of 1997, with a baby on my hip, I had a part-time business curating garden art and antiques and was working full-time as an environmental engineer. My husband and I had just purchased a new home with no garden to speak of. With no time to till, I covered the main bed with 6 inches of composted leaf mulch. When I began to plant, I dug and amended one hole at a time. I realized my busy schedule built a better soil. Simply put, tilling messes with the soil structure and the ecosystem.

Today, my attention turns toward accelerating this natural process by adding [inoculated] fungi into the soil. An added bonus will be a crop of harvestable mushrooms! Going beyond improving the soil's tilth, fungi are also being used to degrade a wide range of soil pollutants, including herbicides. I've selected the native king stropharia (*Stropharia rugosoannulata*) for my garden because it's a good edible and easily identified. It's a beautiful fungus, with dark burgundy-colored caps and white stems.

King stropharia mushrooms are also called garden giants and spores are available from Tradd Cotter's online store, shop.mushroommountain.com.

Watering by Hand

Using the Essential Skill of Observation to Keep Plants Hydrated

HERE I STAND, holding a hose, happy to have fresh water at my fingertips. A landscape truck passes by, jingling with irrigation parts. My neighbor has just updated his watering system. It twirls and streams as the water runs off into a drain, none of it absorbed by the already soaked soil. The system he has installed has complete coverage; it is programmed to turn itself on and off with a timer—no one has to do a thing. I imagine him looking at me and thinking: "What kind of an unsophisticated, homespun garden is that—that guy couldn't even put in a proper irrigation system!" As I happily work out a kink in my hose, I think to myself, "If I had such boring plants, maybe I'd want to hide them in a cloud of water, too."

Watering takes time; it requires critical skills. Many of us remember our grandparents standing in the yard with a hose. We may not have noticed then, but they weren't just squirting it willy-nilly; there was a method to it.

They dragged the hose around and aimed it at plants that needed a certain amount of water, at a specific time of day, in the right season. They would single out and address plants that needed water, even while they were sitting right next to plants that didn't. They were using their watering skills. It's time to give those skills their due respect.

One-size-fits-all watering systems don't address all issues, and at their worst they can create new problems. Poorly used and installed, they're incredibly wasteful, they drown plants, and they create a sense that everything is taken care of, that you don't need to be involved. They sabotage our watering skills and erode our understanding of plants as part of a larger system. We expect expensive systems to make gardening easier and better. We bought them thinking that they'd make our lives more carefree. In the end, they end up controlling us in ways we don't even realize. It's a mistake to think that any automated system can wholly replace watering skills. There

LEFT Seeds of many old-fashioned annuals, such as this pink petunia, will rot under constant irrigation. The exuberance and success of this garden comes from restraint, from not having an irrigation system.

RIGHT My grandmother and nephew, sharing the joy and skills of using a hose.

are certainly fantastic systems, and they are getting better all the time, but over all, rather than improve watering methods, the influence and popularity of automatic systems has slowly started to alter gardening techniques and even had an effect on the selection of plants that are available in nurseries. Whether it's conscious or not, we've begun to select plants that thrive or tolerate frequent sprays of water on their leaves or on saturated soils. Other plants have been edged out and forgotten.

These systems are still very new. The idea that millions of small homes and businesses should have plastic pipe running around and under the lawns, spraying plants, benches, and parking areas with a fine mist of clean water—even when it's raining—is a very recent development. It wasn't until the 1960s, when the increased availability of affordable PVC piping allowed us to install irrigation systems.

Sure, there were irrigated gardens before that. For example, Smiley Burnside was a doctor and a gentleman of the old school. Competitive in the camellia world, he grafted, selected, and entered his flowers in camellia

A Brief History of Water Relations

Irrigation goes back a long, long way. Egypt and Sumatra fed themselves from their rivers as early as 5000 years ago. From ancient civilizations to modern America, misuse of irrigation and the misunderstanding of water, soil, and plant relationships have led to failed economies. The collapse of the Mayan civilization is now blamed on a very mild water shortage. Drought left water levels low, but they were only 25 to 40 percent less than normal. That incredibly advanced city-based civilization disappeared because they didn't pipe in water. Even when water isn't low, problems can compound. Salt accumulation and groundwater pollution can also lead to failed irrigation.

In another more recent example, on a plantation garden near Charleston, South Carolina, a huge percentage of the antique rose collection was wiped out in a matter of months. As the plants declined, the owners watered more. No one bothered to discover that the irrigation system had overdrawn the well. The system was pulling salt water from the nearby marsh. As it watered more and more, it killed the garden. Even during the crisis, the owners wouldn't acknowledge that their irrigation system had delivered the toxins. Finally, plant pathologist and sleuth Kari Whatley, of PlantScout, stated the problem simply: "Overconfidence in the technology and lack of detailed observation combined to allow the irrigation system to constantly deliver contaminated water to the plants." The same can be said of many civilizations, which crumbled because they failed to understand water.

shows around the South. A successful man, Smiley had the resources to hire an English garden designer and to install an irrigation system in the 1940s. It was still there, in the 1990s, when I restored his gardens and wrestled that system. It was like a living thing; a helicopter-like pump sat at the edge of his lake slurping up water. An industrial-sized startup switch in the basement of another building activated miles of galvanized pipe that throbbed and thumped with primal power pushing water up a hill and across 8 acres. It was extravagant for the time, but it wasn't automatic. Even then, with a brand new, state-of-the-art system, Smiley knew you applied water only as needed.

Before I was even aware that some people could afford such extravagances, I watered with my family—my teachers—the same way everyone else watered. We used repaired hoses that kinked or were 10 feet too short; we used buckets that were always in the wrong place. Even old leaky pots and plastic milk jugs that soaked our feet helped us get water where it was needed. Ironically, the house and garden where I grew up were previously fed by a ram. A ram is a very clever, self-powered device that uses the kinetic energy of flowing water to pump a small stream of water a long way. Rams were used in Egypt to move water to the top of the pyramids; they're used today because they're easy to make and they pump water for free. In the nearby swamp, a half-mile away and 100 feet below, remnants of the ram walls and bits of pipe leading to the house still remain. I'd love to put it back together, but it would cost too much in time and money.

I grew up with the understanding that water is precious and needs to be applied to plants at specific times and when certain signs from the plant and the soil communicate to you that it's needed. In the gardens I've established in my professional work, some had irrigation and some did not. Riverbanks Botanical Garden had an expensive system that we scrapped in the first year. Today, the garden does have an amended, simpler system of irrigation. But this, like any great garden, is not subjugated by automation. Successful gardens depend on clean water provided to plants when they need it, by people who understand and embrace the relationships between plants, soil, air, and water.

The Teachers

RYAN GAINEY AND JIM MARTIN

There is a primal pleasure in touching water. I've met so many gardeners who tell me that their favorite thing to do in the garden is water. They'll tell me how much they love to water their rosemary, or stick a hose pipe in the dirt, or use their special galvanized buckets. Watering is nurturing, relaxing, and engaging all at once. Ryan Gainey and Jim Martin, two important men in my life who've encouraged and supported me in different ways, value the skills and pleasures of hand watering.

Ryan Gainey waters his garden by hand, but he takes off his rings before doing so.

Ryan Gainey is an internationally recognized garden designer. His long career involved commutes on the Concorde to design glamour gardens in Europe and across the United States. He was featured in Audrey Hepburn's acclaimed PBS series *Gardens of the World*, and in projects with Atlanta's High Museum of Art. Despite working at these lofty heights, Ryan told me, "I grew up in the country. Poor. The women who taught me to garden—honestly, their names were Miss Flowers, Miss Fowler, and Miss Fells—watered everything with buckets and long-handled dippers and always in the morning, after breakfast chores were done." Ryan is proud of the people and places where he grew up—rural, poor, and sharecropping tobacco farms outside Hartsville, South Carolina. He embraces the styles and lessons he learned from these country gardeners.

Today, in his celebrated Decatur, Georgia, garden, which has been featured on the cover of *House and Garden* and was a set for a Disney movie, Ryan still waters with watering cans. As you enter, the old, galvanized tin cans sit atop a stone wall. They make a cute and quaint display, and they are regularly put to work. This stunning garden is watered by rain, by hose, and by those watering cans that sit front and center, just inside the garden gate. That's where they are handy. And that's where they make a statement: this is a handmade garden.

Ryan says that subordination to automation can kill some plants by drowning them, causes foliage issues, and promotes soil-borne pathogens that wouldn't be able to thrive without the constant water. Ryan, who has rooted miles of boxwoods, once told me, "Never irrigate boxwoods; it will kill them." I'd have to counter that irrigation doesn't exactly *kill* boxwoods, but overwatering encourages pathogens in the soil that can kill them. In a

client's garden, I struggled with a sad hedge of box-woods for years. The problem was rooted in the client's desire to locate this hedge around a lawn. The soil was sticky clay, and the lawn's irrigation systems kept everything very wet. The American boxwoods died. Eventually, I replaced them with Korean 'Wintergreen' boxwood, which can tolerate wet soils better than any other. The substitution worked, but this illustrates how plant selection was dictated by an overused irrigation system. The better solution would have been to simply choose something different from a constantly perfect, thirsty, putting-green lawn paired with a hedge that likes to be a bit dry.

Ryan's garden is built on the ruins of a 1920s nursery and greenhouse. He has photos and records of the operation showing containers everywhere, which was common in early nurseries. We can learn from this. In the time before plastic nursery pots, plants were mostly grown in the ground and sold bare root or later in various pots made of materials such as metal or clay. Clay pots were heavy and messy; they didn't lend themselves to the kind of cash-and-carry nurseries we have today. Even though we don't use clay pots in nurseries anymore, they can add a rustic, earthy feel to gardens and on patios, and plants love them. Understanding the way water and plants relate to clay helps your plants thrive. Clay loses water through every pore. One way to reduce this, in periods of drought, is to plunge the entire pot into the ground. In old nurseries and greenhouses, including the former occupant of Ryan's garden, there were special plunge beds for this. Those beds were filled with leaf litter or ground bark that retained water in summer and provided insulation. Ryan offers a tip for watching moisture levels: "Know the

LEFT Ryan Gainey's watering cans sit by the edge of a fish fountain where they are easy to fill and use in his hand-watered garden.

RIGHT Ryan Gainey's famed Decatur, Georgia, garden relies on the structure of boxwoods, a plant he says should never be under irrigation.

Tin-Can Gardening

In 1932, Walter Studley started the first private nursery in Michigan. Mostly, he was growing tree seedlings to reforest the state. There were no plastic nursery pots. Most nurseries of the time grew trees in beds in the ground, and then the trees were transplanted in the fall or spring. Full of good ideas, he began to grow plants in containers, using discarded tin cans from the Michigan Fruit Canners and Lloyd J. Harris pie factory. Jane Huntree recalls working in the nursery: "We still used those nasty, dangerous cans. We had to be very careful not to cut ourselves on the rusty edges. I remember Howard's picking up a truckload of one gallon and three gallon cans. Some still had food bits in them. That was the day before dump beds on the trucks, so we'd have to stand in a truck of dirty cans and throw the cans into a pile." Another problem that people came up with all sorts of ways to deal with was storage, since you can't really stack empty tin cans. They used special jigs and tools in attempts to crimp the cans to make them stack. Alas, the cans still tumbled, rattled, rusted, and were a pain to get ready for planting. Nevertheless, in our little nursery, people seem to find them quaint, so for special events, I pot things up in recycled cans and often use them as table decorations—and I try to avoid discussing all of their problems.

Old, galvanized nursery containers used as front-door planters.

ring, the tone, the sound of terracotta. Thump a pot and listen. Solid sound means the soil has moisture, while a dry ring means it's time to water." So the sound of and the look of a clay pot is a way to read the moisture content of the soil. A modern nursery manager told me he could do this with plastic, but I could never get a handle on the tones.

There was a time of transition when people grew in repurposed tin cans. And you still can, if you so desire. If you punch a hole in a tin can, it makes a cute, recycled pot. But be warned, water rusts metal. In my experience, it just takes a year to rot a coffee can. You can slow this by choosing low-water plants, such as agaves or succulents. It's also a good idea to place the cans on decks or patios. When set on the ground, the roots will grow through—not only your water holes, but weak spots in the metal—and start growing in the ground. Ever picked up a plant in a tin can, only to leave the plants, roots, and soil behind, all perfectly formed by the can?

"Know the ring, the tone, the sound of terracotta." In old nurseries, there were often special beds for sinking clay pots in rich, moist soils.

For many of the reasons described here, tin and clay were both seen as having inherent problems. So they mostly went away, along with some of the skills and understanding we gained from working with them, and were replaced by the versatile, malleable, and ubiquitous black plastic pots.

No matter what kind of pot you use, the old greenhouse practice of sinking pots works at home, too. Since most of us gardeners have little holding areas, places we keep plants in pots, an easy way to reduce water needs is to sink them, just as they did in the floral nursery that Ryan's garden was built on. In my own holding area, I use a variation on this. I dug holes for several empty containers and installed them in the ground. Those containers act as receptacles, or sleeves, where I can slide other potted plants into them for holding. Since they're insulated in the ground, moisture stays in the pot longer, and the plant doesn't have to be watered daily. It also helps the plant to stand up better, propped up by the sleeve. Many large nurseries have entire fields managed this way, known as a pot-in-pot system.

The most important thing to remember is that your plants' watering needs often change depending on what type of pot you're using. And watering all those pots and plants can be a lot of work. Ryan has an occasional helper and the resources to pay someone—a knowledgeable someone—to water when he's away. The rest of us just have to do the best we can within our budgets. In our garden, I have dozens of metal, plastic, and clay pots that need regular attention, but I don't have the resources to hire additional help. Fortunately, I'm lucky enough to be able to ask my mother to do it. I also group my pots and set up a temporary hose and timer to water while I'm away. But I can see the difference when I get back and return to watering by hand. The little droplets of water from the sprinkler, no matter how long it runs, won't all find their way into the soil. No matter what type of pot you choose, or whether or not you have a helper or a timer or an entire system, when you're available to do it, nothing satisfies your plants' water needs like doing it by hand.

You *can* have a great garden without an irrigation system. Plants *can* thrive. It's good for the garden, and it can be good for you, too. Time spent watering is your meditation time, a time for learning and observing. It keeps you close to the ground, and closer to your plants and their world. But you have to cultivate good watering and observation skills to succeed. And Jim Martin has spent his career showing thousands of people how to do just that.

Jim Martin is a professional horticulturist and farmer, and a really busy man. He's been director of the three major botanical gardens in South Carolina, and he's in charge of the plantings in all of Charleston's public parks. In each, he simultaneously planted, designed, budgeted, raised the funds, and set strategic plans. His powder-blue van, wrapped in a giant sticker that reads, "Compost in My Shoe," is always full of projects, and even a few changes of clothes. He runs an organic farm, a specialty produce business, and keeps bees and a killer home garden. You'd think an irrigation system would save him time. But he, too, does it by hand: "There are many days I come home so tired or bothered and tempted to do nothing. But I see the garden needs water, so I have to do it, and whoosh, I'm absorbed in the garden." Absorbed; connected; communicating with his living plants. Jim asserts:

> I think one of the biggest problems with an irrigation system is that it takes you away from a very intimate, critically important task of gardening. On my farm, I do use drip irrigation, but in my garden, I hand water only. It's my motivator. I have to do it, I love to do it, and I don't see it as a chore. Things are happening that I need to see. When I water, I'm watching what's going on.

Watering by hand helps focus attention. You see things, smell things, and understand the world that exists inside your perennial border.

Another way Jim deals with his watering needs is simply by learning more about his plants and how they adapt to and tolerate stress, and then planting things accordingly. Under the live oak, for example, he has tough shade plants like bromeliads and dwarf palms and intricate trellises holding container plants. In a ditch out front, which floods in big rains, he grows *Hymenocallis* bulbs, around here called Fourth of July lilies, which are riverine bulbs from areas that flood seasonally. Lately, this has been known as hydrozone gardening; simply put, pick the right plant for the right place. Jim adds that we also have to understand that many plants tolerate drought stress just fine, even if they wilt. By way of example he tells me, "You wouldn't believe how tough autumn fern is. I have entire sections full of the plants that can tolerate stress, so in a bad summer, I let them go. You know, even grass can go dormant in the summer. It comes right back."

Still, for all the condemnations I've made of irrigation systems, the truth is that the systems themselves are only part of the problem. We install technology that's billed as something to help us garden better, then we let it drown our plants, waste our water, and cause us endless frustration. We know this; we watch it happen, yet we still leave them running.

ABOVE *Hymenocallis*, also called spider lily, thrive in places that are alternately flooded and bone dry.

BELOW Jim grows vegetables and herbs in a vertical garden watered with a creatively painted drip system.

Rehydration Therapy: The Many Meanings of Wilt

I always groan when I hear someone say, on a broiling July day, "The plants are wilting; we need to water!" That automatic response shows a substantial lack of understanding as to how plants use and conserve water. Whenever you see a wilting plant, your first concern should not be how much water to give it, but to figure out *why* it's wilting. Even in saturated soil, a plant may wilt. On really hot days, a plant may be wilting to protect itself, to conserve water. Since plants lose water through their leaves, the leaves simply close up their little water-release valves. Plants respond to heat in many, many ways, including reducing gas exchanges and stopping water uptake. An experienced gardener recognizes this and will wait until the cooler evening to see if the plants perk back up. Older gardeners use the word *flagging*, which means it's drooping because it's hot and tired. (Note that plant pathologists use the term *flagging* to indicate the death of a leaf or twig, which is often a symptom of disease.) Flagging is sort of prewilting. You can't really tell the difference between wilting and flagging simply by looking at the plant. You have to consider everything else: humidity, recent watering, heat, cloud cover, and soil moisture. In your own garden, you usually know if things are well watered. If you watered well but still note some flagging on a hot afternoon, don't do anything. The plants don't need water; it may even be harmful to them. If all weather conditions are normal, the plant might close its stomata and look wilted for a few hours in the afternoon, but rehydrate in the evening. Watering skills are simply honed with time, and often imperceptibly. Watch other gardeners, ask questions, and make watering skills a part of your conversations; you'll develop a new vocabulary and new tricks. In this way, I often learn tricks that people don't even realize they have because they've taken their own knowledge for granted.

But technology can yet have its place in the garden. New types of systems have evolved toward efficiency in water use and installation. Jim says:

> Drip and surface are great, but overhead usually wastes too much and causes problems. I do use drip on some of my vegetable farms out on other islands. Things happen and I might not get out to those fields for a few days. But when I'm around, I hand water. Even among small-scale, organic farmers, there's this groupthink that we must irrigate, we must drip, we mustn't waste water with a hose. I water to my plants' needs—even in a row of eighty-five chards, I know which ones will dry out first due to some soil changes.

Jim is fascinated with epiphytes—plants like some bromeliads, orchids, and cacti that grow without soil—and vertical gardening. He's constantly testing wooly pockets, shopping carts, and anything that can be hung onto walls or from trees. His quest: make fun planters and pair them with plants that thrive in the material that they're made of—clay, metal, plastic, mesh, or synthetic macramé with appliqués on it. Since the drip system is a necessary part of that, he makes his fun and funky. His drip hose is painted pink and pinned to the wall in spaghetti swirls, and the tiny purple emitter hose spirals out toward the hanging plants. If he has time, he color coordinates the planting and painting of the tubes. Still, he turns the system on as needed.

But for the bulk of his garden, Jim uses his watering skills, his observation skills, and his understanding of the ways that plants respond to their environment. Most automatic irrigation systems put you in the position of relying on a general system to do precise work, preventing you from making careful observation of the plant's needs. One size does *not* fit all. Different plants have different needs, and even the same plant has different needs on different days.

LEFT This coleus has closed its stomata to prevent water loss during the hottest part of the day. It does not need extra water.

RIGHT Since all conditions were normal, the same coleus, without any additional water, has rehydrated during the night.

Updates and Adaptations

The ability to observe the telltale signs of plant water stress is a skill that makes a great gardener—and a great nursery manager. In our field nursery and in the gardens and landscapes that I plan, people often question the lack of complex irrigation systems. I tell them that we can and sometimes do irrigate mechanically, but we ask broader questions. We use an array of skills and techniques dictated by what plants need. The success and validity of our philosophy and practice is easy to see in our field nursery and in gardens we manage for others.

We water our nursery plants to encourage a healthy growth rate, produce healthy plants, and conserve water and the energy needed to pump that water from 350 feet belowground. To be clear, that's different from the goals of most irrigation systems, especially in nurseries, where the goal is to produce maximum growth in minimum time. Our plant products—crinum bulbs—like most living things, are mostly water. I *could* produce a fat, plump bulb more quickly with constant, intensive irrigation. But that bulb would just be water. Instead, the bulbs we produce are dense and packed with carbohydrates; they're more resilient, and perform more successfully in our customers' gardens.

When considering new plants, consider the water needs of that plant and, just as important, how that specific plant was grown. Plants grown in nurseries with constant irrigation need more water in the transition to the ground than plants grown in drier conditions. We grow our plants in the ground so that they have access to normal soil and soil water. We've tried various systems. Because of their wastefulness, one-time-use drip tubes, which are tilled up or thrown away each year, do not interest me. Instead, we use drip tape, which is designed to run in straight lines or have right angle turns. The main supply lines are permanent, and buried for their protection from solar deterioration; they never need replacing. Our drip tapes last four or five years. Some stay in place below mulch; some are stored and pulled out if needed. All are regulated-flow hoses with tiny internal tubes and emitters. These two things are important. The channel helps keep pressure uniform, thereby watering uniformly over long runs, in some cases up to 500 feet. With an old-fashioned soaker hose, you just can't go that far because the water gives out. Also, soaker hoses tend to get blocked areas as a result of tiny salt and dirt particles in the water. The built-in emitter-style hose that we use allows tiny particles to stay suspended in the water and pass through the emitter, preventing clogging. And since the emitter is built inside the tube, we don't have to punch holes or place emitters—they are evenly spaced in the tape.

An important management technique, in drought times, is that we set simple timers to run the system for hours. The outflow is tiny; the goal is to slowly, deeply saturate the soil, not to keep it constantly wet. We water the dirt, the organic matter, the bacteria, and the fungi in the soil, letting those all hold water and slowly release it to the plants.

Good watering technique can also give the gardener benefits beyond irrigation. Besides conserving water, by using drip irrigation, or none at all, we also reduce weed issues. That's because the bulk of our summer weeds must absorb some water in order to begin germination. As the tender new plants form from the seed, if irrigation is spritzing them constantly, they have ideal conditions to keep growing. So you get a higher percentage of germination and a higher success rate in weeds when there is constant

Our in-the-ground nursery, vegetable, and herb gardens get water via drip irrigation only as needed. I can read the signs of water needs for each plant; if the beans and blue salvia show any sign of stress, I know it's time to water.

moisture present. This is true in nurseries and in landscapes and gardens. This means that overwatered gardens require more weeding or, as is the norm, more herbicides.

In my small city garden, I've never had an irrigation system. I'm on a sandhill, a land formation that is exactly what it sounds like, with coarse white sand and sharp drainage. To choose plants, I looked around to see what grew in old gardens in the neighborhood *without* irrigation. I saw and used *Camellia sasanqua*, anise, Carolina jessamine, and yaupon holly. Then I used soaker hoses and cheap timers for two summers to get plants settled in. Today, eighteen years later, I never water. My little lawn area is not a putting green, but it's healthy and attractive. I've only needed to turn to herbicides to deal with a weed problem once or twice over the decades. But just two doors down, my cousin moved in, ripped out all the old plants, installed new plants, rolled out a lawn, and started overwatering. He used to bug me about what to spray for this weed, and how to get rid of that fungus. He's paid good money for that system, and he thinks he has to use it. So instead of addressing the overwatering problem, he just keeps spraying the weeds that are thriving in the constant mist.

In garden design and installation, my approach to irrigation is similar to that in our nursery. Pick the right plants, care for the soil, and water well during initial growth. I use drip tubes, buckets, hoses, and, most importantly, I educate the clients about root and soil interactions. And once they start putting it all into practice, once they start observing, they're thrilled by what they discover and find, those simple truths, those "aha!" moments.

I often turn to this amazing demonstration to teach people about pre-watering and how difficult it is to wet roots. Fill a bucket with water and get a new containerized nursery plant that's ready to go into the garden. Submerge the entire container in the water and watch how long air bubbles out from the potted plant. The bubbles indicate that the roots were *not* wet. New, lightweight, artificial, potting media have been developed to save fuel in shipping, but not necessarily for better plant health. Plants from nursery factories, in light media, will bubble for ages, but will not necessarily settle into the garden well. I presoak every plant that I use in a garden. For soaking, I often include a weak liquid fertilizer or soluble mycorrhizal inoculant to stimulate root growth and insure that if the plant goes through any transplant shock it will have all the nutrients it needs to settle in. Hold the root ball or container under water until it stops bubbling. For 3- or 5-gallon nursery plants, this can be a slow process. For large-scale gardens, I set up 55-gallon drums of water. For huge containerized plants, I've literally had to rig them with sheets of plastic, with the edges tied or taped at the top of the container and filled with water.

Some of my coworkers will make fun of all this extra work, but I felt happily vindicated when I watched Ryan Gainey planting in his garden. As Ryan explains it, "Every plant that I plant is treated in the following way: (1) Soak the plant in a bucket of water; (2) plant the wet plant; (3) water in with a hose or bucket, drenching to collapse soil around the roots; and (4) hand water for a week or so."

If you're doing your planting far from a water source, there's an old, easy trick that I like to use to get them established. I use this for planting things along the fence lines surrounding our nursery, which generally require about 5 gallons of water each week, delivered slowly. To make that happen, I use a nail to make a hole in the side of a 5-gallon plastic bucket. I can then fill that and let it drip out slowly, maintaining good moisture for the soil and roots—you'll only need to do this for the first summer. It's best if you have a lid for the bucket so that the water doesn't slosh around and spill out as your carry it to the plant. If you want to make it even easier to carry, you could just fill the bucket with ice and leave that to melt; I've seen this trick used in urban areas. And it would be wise to mark the bucket—I spray mine with red paint—so as not to accidentally use it for something else. You could, of course, just purchase watering bags that serve the same purpose, but, unlike your buckets, they won't be repurposed, and they'll cost you more.

If you ask the right questions up front about your soil and your climate, there's just no need for an irrigation system. In one project, after the devastation of a category 4 tornado, Tom and I volunteered to landscape the newly rebuilt Beech Island Fire Department. As we worked with the firefighters to install the new landscape, I kept hearing questions about when the irrigation system was being put in. It never was. Instead, in hard-packed clay, we pushed aside construction trash and picked out holes with mattocks, and then we planted the right plants. In our climate, that included palmettos, redbud, cypress trees, spartina grasses, crinum lilies, rose of Sharon, and Bermuda turf. For the first summer, the firefighters used their hoses to get the plants established and keep them happy, but since then everything has thrived without the help of an irrigation system.

Learning and using great watering skills can save lots of plastic, batteries, computers, and, most importantly, water. Water is the limiting factor in the lives and health of so many people. We simply cannot afford to waste it. When I drive by a poorly managed watering system, spraying clean water during a rain storm, or running into a stream onto the roadway, I'm sometimes brought to tears thinking of my friends in the Caribbean who have no running water in their entire neighborhood. Millions need that water to cook, brush their teeth, or mix formula for a baby. This is a worldwide issue, a looming moral and conservation crisis. As gardeners, as workers of the earth, we must be ambassadors and teachers for these issues. Watering by hand helps us immerse ourselves in the plants' world and learn more about the cycles, pests, smells, and needs of plants and of other people.

Watering-In

While you won't find it mentioned in gardening handbooks, "watering-in" is a phrase that almost every gardener says and understands and expects the same of those that they work with. It's universal gardener speak; I have friends from across the United States and Europe that use it. Where did *you* first hear it? From what tiny moments, conversations, mentors, or plants did you first get your head around it?

Watering-in for a gardener is sort of like bedside manner for a doctor—it's a practice, a gentle, caring, comforting thing. It's making sure that someone who's a little weak or in a moment of transition is tucked in and going to sleep well.

The process of watering-in a new planting, versus simply squirting some water, demands an understanding and appreciation of the needs of roots and dirt, potting soils and water, photosynthesis and positive energy. It doesn't mean that we know everything there is to know about all of this, but there's a feeling.

When you water-in, you:

1. Saturate the root ball, insuring that the plant itself will have water, that everything surrounding young roots will stay wet once below-ground.

2. Provide good contact between roots and dirt. Water moves dirt particles to fill in pockets of air under, around, and in the root ball. These might be little pockets you can't see, or if you're planting a big root ball, big caverns that not only keep roots dry, but will collapse, causing your plant to sink later.

3. Wet the dirt around the new plant, so that the dirt doesn't act like a sponge, pulling water away from the new roots.

4. Wash dirt and mulch off of the leaves, so the new plant can hold it's leaves turgid, reaching for the sun, and photosynthesize.

5. Give yourself a minute to evaluate your work, think a positive thought, or have a Zen moment.

6. Provide humidity around the leaves, encouraging stomata to open and exchange gasses, which will help them to get back to absorbing nutrients.

Collect and focus the energy of moving water in the soil and air around a plant. You might call it chi, positive energy, lining-up, or paramagnetic force—whatever you call it, it pulls together your own energy with that of moving water, plants, and life in soil.

Those simple actions and involuntary connections make life rich. One tiny action can set off a chain of scenes in our minds. Sometimes during a watering conversation, I'll hear in my own voice an inflection, a tiny change of tone when I'm getting excited. I'll then recall an afternoon years ago, on a road trip with a friend, looking over a vast desert, my friend fixated, holding my shoulder, imploring me, saying, "Now? Now, you must be excited! Say it out loud!" Or when I water with a coffee can, I see the smooth twisting of water becoming a muddy stream of cypress pond water, pouring from the bottom of a tiny tin that my father picked up to nurse along a newly planted pecan tree seedling behind a barn that he dreamt of renovating, of making into our house.

Watering-in does all of that for me. It's so elemental, something that builds unforgettable connections. When you teach someone to water-in, make sure it's a fun experience, an important moment; it may be a moment they associate with watering for the rest of their life.

Rooting in the Ground

Working with Pass-Along Plants

*I*N MY YEARS OF SHARING plants, I've traded plants with older gardeners who gave me things they'd rooted and repotted into milk jugs,
paper bags, newspaper rolls, and soup cans. They root at home and
then find a container to share the plants in. Some of them are obsessed by
the thrill of rooting. Some want plants to share. Some simply wouldn't think
of actually paying for a plant, much less a pot, and just can't bear to let a
potentially free plant get away. Learning to root at home may offer any of
these benefits for you. Importantly for all of us, it reduces the environmental
cost of producing, recycling, or sending tons of plastic pots to the landfill.
Rooting at home can be big-picture stewardship, but it can also be an inti-
mate nurturing of a little living thing. It can be a soulful connection—stories and memories passed between people and plants.

For other gardeners, rooting is the only way to get plants that are not
available in nurseries. Quite simply, if you want certain plants, you have to

root them. Otherwise, your garden is doomed to be shrubbed only with the limited selection of plants that nurseries and plant promoters put in front of you. Granted, there are many of us in that very business with great taste in plants, but, even if you know its name, you'll never be able to go out and buy your grandmother's rose—it's just not hers.

Most plants want to root. It's part of their physiology, a way of reproduction. Sometimes it even happens by accident, like when you stick a bit of rosemary in a glass of water to keep it fresh and all of a sudden you see tiny white roots. Rooting in water is easy, in fact, but getting that rooted plant to transfer to the dirt can be challenging. In water, the fine white root hairs are perfectly filled with water, but when you transfer them, those hairs are damaged and traumatized to such a degree that the new plant often dies in the process. Still, there are hundreds of different shrubs, perennials, and trees that *will* root right in the ground. People have been taking

LEFT Cuttings stuck directly into plastic bags are ready to move or give away. While not the best solution, plastic bag pots do use less resources in manufacturing and in transport than thick-walled plastic pots.

RIGHT A classic example of recycling. Tire planters make summer homes for large house plants.

ABOVE My mother's garden is made of up plants she's rooted herself, often from the gardens of friends, relatives, and mentors. Perennials, shrubs, vines, and fruit trees remind her of those people and their connections.

BELOW Cuttings stuck in the sand can be transplanted to pots or moved directly to the garden. A bare twig of a shrub can be a new plant in less than a year; a perennial can transform itself in weeks.

advantage of this for ages, and we've learned many workarounds for plants that are a bit more difficult to root. I have friends who embrace the challenge, who've built entire professional careers on figuring out how to root impossible-to-propagate plants. It's hard not to admire a person who spends their free time hunting down and naming new rhododendron species in wild, virtual deserts, and then figures out how to root them using humus collected from the same woods. It takes an amazing passion for that work; an impressive level of devotion. For the rest of us, though, there are many, many more plants that will root easily. I tend to focus on those in my work.

A few simple lessons will open a world of new plants—and an endless supply of free plants—and maybe even change the way you design your garden. Unlike a lot of modern gardens assembled from nursery shopping trips with one of this and one of that, gardeners who root tend to have more repetition in their gardens. Repetition provides beautiful, romantic continuity to gardens; it's something that both garden designers and visitors love. But the most important lesson to start with is to realize that it doesn't have to be difficult; there is very little investment—just a few snips and squirts. Take ten cuttings and don't worry if only two root; just enjoy the successes.

The Teachers

RUTH KNOPF AND GLORIA FARMER

Two women taught me to root in the ground: my mother, Gloria Farmer, and my friend and mentor, Ruth Knopf. Both are in their seventies, and both have been rooting since they were little girls. They've made amazing gardens with homegrown, handcrafted plants. Compelled by the magic of rooting, of helping and watching that little cutting grow into a beautiful plant, they can still recall the women who first taught them to root. It's a secret of life that keeps getting passed down, in moments when a mentor and a child share in the spirit that's contained in all living things. Eventually, that child roots her own, and the next year, her yard is filled up with plants that were given or broken off in that exchange. Soon, she's getting a cutting here and there, and her rooting bed is overflowing, and

My mother, Gloria Farmer, passes it down to her granddaughter.

she's taking plants to the church, the fire station, and even showing up at someone else's yard sale with a few hydrangea that need a home.

Ruth and Gloria root reminders. Walking through their gardens, they recall Urbana, who shared her "chicken rose," or reputed bootlegger Greg's appropriately named "bootlegger rose." In a way, they root little companions, alms of the people from whom the cuttings came. Growing up in tighter times, regular visits to the nursery to populate their gardens weren't an option, and if they were, they might not have had gardens full of such time- and climate-tested, tough plants, anyway. Rooting lets you take a bit of certainty—an edited lesson from a knowledgeable gardener who wants you to succeed—home with you. I can hear both of them visiting their friends' gardens and saying, "I think I have a place for that, may I take a little cutting?"

Ruth and Gloria root things in designated rooting beds with the idea that they'll move them to the garden later. Their processes differ slightly:

ROOTING IN THE GROUND 99

Reduce, Reuse, and Recycle Your Pots

Indisputably convenient, plastic gardening pots, which rose to popularity in the 1970s, spawned a whole nursery industry because they carry and ship easily. They were an amazing innovation that stimulated the gardening industry and gave many home gardeners a way to get back in touch with the earth, and each other (they made sharing easier, too). Over the years, however, they've been reinforced with another layer of plastic, and come sporting logos and trademarks. While old black pots could be used over and over, many of today's pots, by order of the trademark holder and marketer, cannot be reused by nurseries. Of course, there are recycling efforts, as well as new, interesting, and more sustainable pots. I grow plants in pots made from compressed cow manure. There are even pots made of living fungi—mushrooms enticed to grow into molds. Still, rooting at home, in the ground, simply reduces the need for any pots. Reduce, reuse, recycle, yes, but it's also time to rethink.

Ruth roots in the summer, Gloria in the winter. Other people do things in totally different ways: rooting by a spigot, by the compost pile, or rooting right in the place where the plant will grow for the rest of its life. In other parts of the world, people stick stripped stems, that will become woven fences, sculptures, and flowing masses of soil-stabilizing groundcover.

Ruth Knopf was a preacher's wife. She traveled, cooked, and busied herself with the church and her husband. Her personal loves were her family, horses, and roses. I met her much later in her life, after she had redefined herself as a rosarian, internationally recognized for her work in preservation and design with antique roses. From a beach bungalow on the quiet end of Sullivan's Island, South Carolina, Ruth's passion led her to spot long-lost roses in country gardens—in the South, in the West, in Bermuda, in Mexico, and just about anywhere you can name in Europe. Quiet and unpretentious, Ruth has given lectures from New York to Los Angeles, telling her stories of collecting old roses and saving them from extinction. A mutual friend, a professional horticulturist, told me not long ago, "When I water my garden, I think of Ruth's trip to Bermuda and Ruth's trip to Orlando; every rose is a cutting from Ruth's travels."

Ruth taught me that while most people are always saying you should root in the winter, whatever the season, you'd best seize the moment because you may never come this way again. If you don't act fast, that rose might soon be paved over by a parking lot or "cleaned up" by a well-meaning fella with a weed eater. And in the summer, of course, you can actually see the flower, so you know what you're getting. Over the years, Ruth has perfected a surefire way of taking summer cuttings of roses and shared her methods with many people. Fortunately, her lessons will work on lots of other shrubs, too, from azaleas to weigelas.

She's found that the perfect time to take a cutting is a few weeks after flowering. Find a flowering stem with a faded flower or, better yet, with a fruit that's just starting to form. Follow down the stem about five inches and bend it there. If the stem bends and is completely flexible, it's not quite ready. And if it snaps off cleanly, completely, it's past ready. However, if it bends just a little and one side breaks, or sort of splinters, revealing whitish, green wood—that's what you're looking for. That will make a great summer rose cutting.

Ruth speaks from decades of experience rooting roses that are often unique. She'll knock on any door, anywhere, and ask about a plant, though she rarely asks *directly* for a cutting: "We were just riding by and saw your beautiful flowers. We just want to say how beautiful it all is. Could we take a picture? What do you call that?" Soon she's in a deep conversation and invited to have a bit of anything in the garden. This isn't manipulation, though; it's just her good manners. Ruth learned from experience that when you come across something special, or even something you simply haven't seen before, you'd better get a cutting. In her lectures, Ruth always tells the sad story of Mr. Willie. Driving the backroads of South Carolina, she spotted an old man on his porch, behind a yard full of flowers. This was Mr. Willie, smiling, and waving to passersby. She stopped to say hello, and walked through the yard with him, and just before she left, asked if she could take his picture. He said, "People take pictures of my flowers. No one ever asked to take a picture of me." Months later, Ruth went back to visit—and for cuttings—only to find that Mr. Willie had died, and his flowers had been cleared away.

I have an aster I rooted from Ruth, which she rooted from Fanny, her housekeeper in the 1970s. That aster thrives in the South's humidity,

Ruth holds a handful of cuttings, showing that the perfect time to root roses in the summer is just as the hips (top) begin to form.

Any old barn, house, or graveyard is a reason to stop to search for tough old roses and bulbs while touring the backroads.

when almost all other garden asters rot. Fanny's aster is a spectacular plant for many reasons. Its masses of lavender in October are layered with my memories of Ruth giggling and telling me stories about Fanny and teaching me that purple flowers in the garden soften the transition between other colors.

Ruth took me under her wing—or, to be more precise, into her gold, 1979 El Camino Conquista—and we drove through backroads peeking into old gardens, roadside shacks, and cemeteries looking for roses to root. We had the most fun in the country gardens of old ladies and gentlemen, like Bennett Baxely, a man who grew up scavenging plants, and who would became a mentor and friend to me for more than twenty years. We made lots of these southern visits, and we filled up that El Camino with scavenged plants and bags full of cuttings. Our travels took us even further south; I'll never forget arriving late one night in Tequila, Mexico, just Ruth and me in a tiny car full of cuttings and cameras. Much to our surprise, the Festival of the Virgin of Guadalupe ensured that every room was booked and the town was packed with bands and revelers. Finally, we found a single, concrete room with bunk beds, no screens, and a bathroom two floors down. Fortunately, it did have a sink that we could use for soaking our cuttings. Between the marching bands, the celebratory gunfire, and the

What's With Willow Water?

It was Ruth's friend, Malcolm, who first told me about willow water; I figured it was a myth. He told me to take a few willow cuttings and chop them into inch-long bits, then put them in in a jar of water overnight, making a weak tea (the technical term for this is *diffusate*). The next day, soak your rose cuttings in water for a few hours. This is said to stimulate rooting, though modern researchers debate the effectiveness of this process. Willow water contains a hormone called an auxin, which is water soluble and can stimulate rooting; some research even suggests this can prevent some soil-borne pathogens that may infect the cuttings over the next few weeks. Some argue that if you really want to stimulate rooting, you should buy a ready-made auxin compound—a rooting compound containing IBA (indo-3-butyric acid) is effective. But, to be honest, I rarely use the stuff. Mine often ends up neglected on a shelf, absorbing humidity and turning into a clump to be thrown away years later. And, as I said earlier, why shoot for a perfect rooting record? If you stick ten and only five root, you've lost very little and still gained five new plants.

But whatever its success rate, my take on techniques like willow water is that there is a lot more to them than just science. For example, every time I make it, I think back to Malcolm telling me about it. Again, there is that soulful connection, a sense of tradition, memory, and satisfaction. By engaging in these cost-free experiments, we are bringing ourselves closer to our plants and our past.

roosters, we barely slept and spent all night giggling to each other—this was real rose rustlin.'

Of course, rose cuttings will certainly root in the winter, too. That's how my mother does it. She's had a little rooting bed for years. When I asked her where she learned, she fondly recalls growing up poor and happy in the swamps of Crocketville, South Carolina:

> I remember grandmother rooting camellias and azaleas in the ground. She had a little spot, and she'd put mason jars over them. Like the glass cloches of wealthier gardeners, the glass keeps the humidity up and encourages rooting. I've seen people use plastic bottles, bags, or even food-container covers—anything mostly transparent works. She'd root things in the spring and summer, probably when they were blooming. But I root lots in the winter.

LEFT Gloria uses old cake and cheese plate covers to hold humidity around cuttings.

RIGHT Roses (upper left,) figs (middle), and rose of Sharon (upper right) cuttings.

Professional horticulturists and nursery managers, right up until the 1970s, commonly used winter, in-the-ground rooting. Even in the early 1980s, in my nursery and propagation course at Clemson University, we studied how winter cuttings work. The best time for taking winter cuttings differs with your climate. Dormancy is key—otherwise, the cuttings may try to leaf out or they may use energy for shoots that they can't support, rather than roots. Start winter cuttings anytime after the plants go dormant. "I usually start them when the pecans fall in October or November," says Gloria. Another way to remember to do it is to stick the cuttings in the ground or in your rooting bed when you plant pansies. Some people simply stick cuttings in with the pansies, directly in the ground. In late spring, when the pansies need to come out, the rooted cuttings are ready to move. In places with colder winters, you simply root in the spring and through the summer.

When you take a pencil-size cutting—any cutting, from anywhere—if it's healthy, the wood will first start to heal itself. Try to make a sharp angled cut at the bottom, near a leaf bud. Then stick it in the dirt; as it sits there all winter, a callus forms over the bottom, cut end. Later, as the ground temperature changes, roots start to grow from that callus. However, this all depends on having soil that hasn't frozen solid. In climates where the soil *does* freeze, you'll

After just two months, callus tissue forms and a few early roots, too.

mostly follow the same process, but you'll want to bundle the cuttings and lay them horizontally and bury them in deep sand, which doesn't freeze; there they will begin to form that callus. Then, you stick them in

Root-in-the-Ground Woody Plants

SCIENTIFIC NAME	COMMON NAME	MY FAVORITE VARIETIES	EASE OF ROOTING
Aloysia virgata	almond bush		easy in summer
Berchemia scandens	supplejack		moderate in winter
Baccharis	saltbush		easy in winter
Catalpa	bait tree	'Purpurea'	easy in winter
Campsis grandiflora	Chinese cow itch (trumpet vine)	'Morning Calm'	moderate to difficult
Ficus carica	fig	'LSU Purple'	easy in winter
Forsythia koreana	yellow bells	'Illwang'	easy in winter
Morus	mulberry	'Contorta', 'Unryu'	easy in winter
Prunus persica	dwarf peach	'NC State Dwarf Red'	easy in winter
Rosa	rose	'Crepuscule', 'Sombreuil'	easy in summer or winter
Sambucus nigra	elderberry	'Lanciniata'	easy in winter
Vitis	muscadine	'Black Beauty', 'Black Thomas'	easy in winter

Root-in-the-Ground Perennials

SCIENTIFIC NAME	COMMON NAME	MY FAVORITE VARIETIES	EASE OF ROOTING
Anisacanthus quadrifidus	hummingbird bush		easy
Symphyotrichum oblongifolium	southern aster	'Fanny'	easy
Bacopa caroliniana	Carolina hyssop		easy
Chrysanthemum	garden mum	'Miss Gloria's Thanksgiving Day', 'Button Yellow'	easy
Pennisetum purpureum	elephant grass	'Oceanside'	moderate
Salvia	salvia	'Jenks Farmer', 'Henry Duelberg'	easy
Tagetes lemmonii	perennial marigold	'Compacta'	easy

Fig cutting rooted after two months in the ground.

the spring. Anywhere you do this, and everytime you do this, some cuttings will fail. In the South and the West, for example, a warm winter cuts into the survival rate, as the cuttings can grow leaves rather than roots. Other things can go wrong, too, such as rotting or rodent damage. But if you start out with plants that are easy to root, even if you lose a few, you'll have tons of plants in just a few years.

By spring, this rooted cutting has become its own plant and will start to leaf out. Many professionals suggest leaving that new plant in place until midsummer, but I often move things in the spring because the days are cooler, there's more rain, and they settle in more easily. Gloria is more laid back about the timing: "I move them when they seem ready and when I have time." What she means by "when they seem ready" is that when she pulls on the cuttings and they don't pull up easily, she'll dig one to evaluate their eager white roots.

One of her mentors, Urbana Vaughn, taught her some rooting tricks that have been passed on for several generations. Urbana often shared a rose she called the chicken rose. As a little girl, in the 1930s, she and her sister were in charge of feeding and cleaning the chicken yard. This included the process of burying dead chickens. To mark the graves, they stuck cuttings of a pink climbing rose; forever after, 'Climbing Old Blush' was known as

A Rooter's Plea for Chaos

There's one more method for easily rooting things. Gloria recalls how "Old Dr. Mole" rooted azaleas. He'd just bend down an azalea branch, put a brick on it, and root it that way. Under a brick, the dark, moist soil stimulates rooting along the branch. The "mother" plant continues to provide nutrition and water, but within months, the branch can be cut off at the brick, yielding a new plant. This is one of those tricks learned with time, when the azaleas get big and overgrown. If you let things go a bit, you don't even have to use the brick. Florida anise, jasmines, indigo, and yellow bells all become ground-covering masses as their branches touch and root, but you have to *let* them. If you come along with an edger or shears on a regular basis, keeping things overly manicured, you'll miss the magic—and the chance to learn, to discover that some plants spontaneously root. Let plants thrive, spill, and ramble, and many will soon root in the ground on their own.

the chicken rose. Urbana rooted plants the same way that her mother did, right in the ground. She showed us her method of putting a cutting in the ground at a slight angle, then stomping down on the ground right beside it. And she does all of this in the same place that her mother did, a shady bit of open ground right at the edge of the house, where rainwater from the roof seeps over the plants.

Walking through her own garden, Gloria points out the wavy leaves of the unique noisette rose that she rooted from Urbana, along with many other shrubs that tell stories of old friends. A fig tree with huge purple fruits came from Bill Adams; a pearl bush, in full flower, from Urbana; roses from her sister-in-law and from the old man she used to take meals to for Meals on Wheels; and lots and lots of plants that just say "Ruth's" on the tag. 'Mrs. Henrietta Washington' rose came from another lady, about which she confides, "I lost that one once, but since I had given cuttings to Riverbanks Botanical Garden, when I lost my plant, I got cuttings and have Henrietta's rose again."

Gloria's advice is endless. She says, "Regular hydrangeas are easy, but oakleaf is a bit more difficult. With oakleaf, you can do it in the summer but be sure to take all the leaves off." She does woody perennials this way, too: chrysanthemum, salvia, perennial marigold, phlox, almond bush, and butterfly bush. Sure, she sometimes buys new plants, as well—but only *one* of each plant. Then she gets to work, often through trial and error, making cuttings of that new plant. She experiments, and then shares with friends who also root; this is her learning and sharing process.

She roots so much, in fact, that it becomes a constant dilemma in our little nursery on the farm: the rooting bed, which Gloria and I share, is always overflowing. I root a few special things for clients, things that other nurseries just don't offer, or things I want to use in a garden design in such numbers that it would otherwise be cost prohibitive. But for me it's mostly a hobby bed, a place to save the cutting someone gave me as a means of remembering a trip, a garden, or a friend. It ends up being really diverse as a result: there are red roses from China, Imperial chrysanthemum from Japan, salvia from Mexico, and rose of Sharon from a cemetery on Long Island. But Gloria's plants usually crowd out mine. She pots them up and plants what she can find room for. When she's out of room, she wonders if she could plant one at her friend's, her sister's, and her rental house. For the most part, though, it's the process that matters most to her. She'll load potted plants into her truck and give them away at the yard sales and fundraisers for the local historical society. Then the rooting bed is cleared, fresh sand is added, little rows are lined up, and the cycle of our rooting life continues.

——————————Updates and Adaptations——————————

When I told him that I don't know any one person in particular who inspired me to use a variation on rooting in the ground, Felder Rushing, horticulturist and southern garden expert, told me, "You were just raised right." Instead of sticking cuttings into a rooting bed, I do so in the garden to create a quick mass. This is a great money-saving technique that's been used in gardens through the ages. Where the gardener knows how to root, there is a nice consistency, a repetition to the plants, rather than a garden of single-specimen plants. Rooters simply can't throw things away, so they get plant repetition by default. I don't like seeing gardens where you can count how many pots the gardener just bought. So, for mass, repetition, and maturity, when I plant new plants, and new gardens, I often break off bits, stick them in the ground around the original plant.

The most common use of direct cuttings is with willow hedges. It was common in English gardening, often associated with willow basket making, but also seen in New England and the Northwest. I first saw it done in a garden in Vancouver, BC, then later in beautiful, tidy allotment gardens outside of Copenhagen, Denmark. They call these little spaces *Kolonihavers*, or colony gardens, because they are gathering places, mini retreats and colonies where the gardener can attend to her gardening bug and be surrounded by others who share their obsession.

Contorted mulberry limbs, painted and stuck in pots for winter decor, will root in, leaf out, and grow in just a few months. These are from Riverbanks Botanical Garden in South Carolina.

For willow hedges, use dormant, 1-inch-thick willow stems, up to 4 feet long. Bury them halfway, usually at an angle, in a line. They callus and root over the winter, then start to grow in the spring. As they grow, you twine the flimsy stems together to create a woven, tight fence. Some people get creative and make arches or designs. This artful fencing works best in climates with shorter, cooler summers than anywhere I've ever gardened, because in warm places, the willow roots and then grows out of control, requiring weekly pruning.

I take this on a slightly different track. In new gardens or if I'm gardening with a new plant that is hard to come by, I use a variation of the technique to create masses. In one garden, I planted a single, small plant of cut-leaf elderberry, which has been difficult to obtain in the South, since most *Sambucus nigra* cultivars don't thrive in that climate, but there is a cultivar called 'Lanciniata' that does. It stays small and provides really light shade, so I love to use it over small perennials. In this garden, I planted one and waited until Thanksgiving; then I took 6-inch cuttings and stuck them directly in the ground, covering 10 square feet. The mass of elegant, lacy leaves now makes a filigree cover over white-top sedge and purple Stokes' aster. This was an easy, cheap way to make a beautiful garden with a rare plant.

Parasol tree rooted in place from 2-foot-tall cuttings.

Plants like willows and elderberry root easily, as we would expect, but I've also used the same technique with figs, mulberry, catalpa, cestrum, and parasol tree. I usually take 10- to 12-inch cuttings, but I learned by accident that you could do the same with really big cuttings. One winter, my dog, Jack, kept digging up everything that I was mulching. The mulch had chicken manure in it, and he'd gobble it all up, digging up little plants and leaving them to die in the process. To stop the massacre, I had to devise an attractive, effective fence. I cut 3-foot long, 2-inch round stakes of Chinese parasol tree. I stuck the beautiful, waxy green stems all around my beds, deep in the sandy soil. A dozen years later, long after Jack could dig up anything, I still have a hedge of parasol trees in that spot. As a tribute, when I buried him, I marked his grave with that same Chinese parasol tree.

People root things in so many different ways; I've heard so many success stories. Once you learn, once you get that obsession, you can produce thousands of plants. You can do it in the summer or in the winter, and in dirt water, Jello, or floral foam. I remember an old camellia guy who had his rooting beds perched right by a stream so the moisture from the soil kept the humidity up; there was so much good energy in the air, swept in above the flowing water. Many old gardens lined with miles of boxwood hedges got lined with boxwood cuttings first. If it's a look you like, learn to root—you'll save thousands of dollars and have extra plants to give away. It's a joyous revelation. We can make life, we can promote it, be a part of it; it never goes away. Every time I shear, prune, or break a branch, I want to save it, to stick it, to share it, and to know that the fragile, twig will live on to be a tree or a shrub—something with a life of its own.

Felder Rushing and Rooting Pass-Along Plants at Home

Felder Rushing's books and lectures have inspired thousands of us to take a more relaxed approach to our gardening, as evidenced by the title of his recent book, Slow Gardening. He recalls his great grandmother rooting in the ground:

Fresh new roots on a recently "stuck" shrub. This plant could be potted up to give away or simply moved straight into the garden.

I cherish around forty different heirloom roses, most rooted from cuttings taken off shrubs found in rural cemeteries and old home sites. All, including The Fairy and Mutablis (the old "butterfly" rose), are repeat bloomers, and many are spicy fragrant. All are sturdy enough to survive for decades with little or no care at all.

I root them like a gardener, the way my great-grandmother Pearl showed me—not production oriented and high-tech like I learned in college horticulture classes. Pearl also taught me that I didn't have to look for a five-leaflet leaf to know where to prune roses, that they would bloom just fine when simply whacked back to whatever size we wanted, whenever needed.

"Proper rosarians" root rose cuttings dipped into rooting hormones, inserting them in good soil and covering them in winter with clear plastic or glass to protect from freezes. But Pearl was a no-nonsense gardener who had me simply wait until sometime after autumn's first frost to stick pencil-size cuttings partway into ordinary garden soil that's been amended with fresh compost, and they would be rooted by spring. This works also in pots of garden loam enriched with compost.

Because only about half of what we put in the ground actually rooted, I learned to simply stick twice as many cuttings as I want to root. Any extras are lagniappe to share with friends and neighbors.

So every winter now, I make flowerbeds do double duty, by pushing rose cuttings into soft soil between cold-weather-loving pansies or violas. Whatever it takes to keep the flowers growing well is good for the cuttings, too. By late winter their cut ends will be covered with knobby

white callus tissue that sprouts into new roots just about when it's time to replace winter flowers with summer stuff. I transplant the rooted cuttings into pots where they remain until time to be sent to their final garden homes.

There is no question that the rules are there to help us be successful. But gardening can also benefit from thinking outside the box, and there's something to be said for less fuss and worry. Pearl's methods are tried and true, and they have worked for centuries. Pass it on.

You don't need a greenhouse, chemicals, or special potting soil to root plants. These roses are rooting outside and potted in Mississippi mud.

Saving Seeds

Treasuring Heirlooms for Genetics and Nutrients

PLANTS JUST GROW BETTER when you start them yourself from seed. Seeds find their way into cracks and crannies, then grow into flowing masses, weaving the world together in ways we gardeners could never, ever coerce any potted plant into doing. I cherish a strappy old crinum that seeded in at the bottom of a cylindrical cactus, throwing its arms and leaves into a big embrace of its spiny friend. Almost every gardener I know has a shared memory of grandparents who grew all their food from carefully selected vegetable seeds. Some have family heirlooms like a tomato seed brought over from Poland and passed down. Some have family traditions, like winter days spent coordinating who gets what from the seed catalogs. It's a weird state of affairs that growing things from seed today has become a rarity; not too long ago, it was the way everybody *started* gardening. What were the first seeds you collected and planted? Mine were marigold seeds,

"stolen" from the nuns' garden at my elementary school. I'd fill up my lunch box and sow the seeds along with daddy's germinating squash.

Older gardeners, by either necessity or philosophy, tend to garden from seed. Many of my mentors share the seed obsession, which bridges all plant specialties, be they vegetable, camellia, cycad, or zinnia lovers. They cherish loquat, camellias, or lime trees grown from seeds saved from making gin and tonics, and they save little pots for things yet to germinate. They have bags of peas in the fridge, stray beans in their key dishes, and a big, hard, black thing rolling around the floorboard of the car—something they picked up outside a motel in Texas. They save seeds, money, and memories with those plants. They start whole gardens from seed or let whole gardens seed themselves in.

My grandfather had a garden—two, really. All around the perimeter were azaleas, gardenias, a huge pomegranate tree, and a palmetto he'd grown from seed picked up on the beaches of Hunting Island, South Carolina. In one sunny oval bed in the middle, mixed together, he grew summer vegetables and flowers. He taught me to save annual red salvia in a paper sack—just cut off the old flower heads, drop the whole thing in, wait a few days, and shake. You're left with a bag of leather brown seed on the bottom and easy-to-discard trash on the top. It's all so simple, and free. Still, it never

fails to amaze me how many people I meet, professionals included, who have either never seen anyone do this, or are surprised and charmed that I do. I often do it in whatever bag is lying around in the floorboard of the truck. In the process, I am recycling, prepping for next season, and reviewing memories of my bald, old "Grangran" all at the same time. Sweet, isn't it?

Seeding in and letting things go to seed helps us recognize the larger cycles of plant life; you may even get a wonderful new plant due to genetic variation that comes with sexual reproduction. And, as a bonus, it sure cuts out a lot of the compulsive work of deadheading. In fact, with some seed-in plants, the most important thing you can do is to just give up any idea of control. Larkspur can really be a pain, coming up

and covering even midsize roses; Jewels of Opar sprouts in the cracks between bricks; and you can hardly get rid of four-o'clocks. Yet I love them all as much as goldfinches love crepe myrtle seeds in the winter. Even if they make too many of themselves, I'd rather garden by pulling out, by editing, than by adding new little plants that need encouragement to fill in.

The cycle, however, depends on getting a good start. Start with great seeds, and you'll have good plants forever. Collecting seeds from your friends and neighbors works great, so if you see something you like, save the seeds. But if you're going to buy seeds, be sure that they're great seeds—though, getting your hands on them is a bit harder than you might imagine. With flowers, it's easy enough, though I suppose not as crucial; but with our vegetables, political history, modern science, and profitability have influenced what you can get ahold of. As one of our teachers describes, regulations led by the singular motive of profit left home vegetable gardeners with little choice in quality seeds.

Due to changes in our eating and nutrition habits and bodies, it's critical that you start with and always use great vegetable seed. One great way to do that is to buy flower, tree, or vegetable seeds from seed saver catalogs. It's always fun to get them in the mail every winter, and you can buy seeds that carry on the stories and histories of their collectors. But with a little work and time, and by applying some simple methods, saving your own seed results in your own seed strains—you can write your own stories, associating plants with your own family, friends, or special places from your past. Gardening from seed gives us control, and as the teachers in this chapter will tell us, we need to be aware of the quality of flowers, veggies, and all kinds of seeds in your garden. Gardening from seed is essential and efficient for veggie gardening, and it's a soulful recognition of the cycles of life.

The Teachers

ROY OGLE AND DAVID BRADSHAW

Where do new vegetable seeds come from? For every seed you buy, some farmer had to have a field of plants and care for and harvest those seeds. Seed breeding is a huge industry that is constantly embroiled in politics and drama. For sixty years, Roy Ogle and David Bradshaw, both now retired from Clemson University, have been in the middle of it all. The industry has changed drastically over the past few decades. From a homeowner standpoint, the entire vegetable seed industry is sort of frozen in time; they've basically stopped making new seed varieties, and the choices of old ones get fewer and fewer. Without going to specialty sources, you literally cannot get the quality or the variety of seeds that our grandparents bought at the feed and seed store. Since the late 1960s, changes in genetic manipulations, public policy, and politics have made it really difficult for home gardeners to buy improved seeds.

"If you buy a little seed pack at a big retail store, you just don't know what you're getting inside," says Dr. Roy Ogle, and he should know. His achievements in classic seed breeding include not just one, but *every* common southern pea that we eat today. And if you eat a sweet potato in the South, it's all thanks to him. Dr. Ogle developed 'Cherokee' sweet potato, the forty-year standard for agriculture and cooking. He spent ten years working with pathologists to breed the perfect yam—disease resistant, tasty, easy to propagate, grow, store, and transport. This is a man who was part of the generation that set the standard for the vegetables that we eat today. He and his colleagues' golden days were in the 1950s and early 60s. Working in fields and labs with taste testers, nutritionists, and plant and health scientists from

around the world, they bred every common vegetable that we eat today. They studied plants from all over the world. They'd find, say, a really sweet potato in CuyoCuyo, Peru, and they'd beg for, borrow, or just steal it. Then with little paintbrushes, bags, and real plant sex, they'd cross that potato with a heat-tolerant one from a garden in Miami. After years of fieldwork, they'd have a better sweet potato for your grandparents' farm and table.

But two things changed, making them the last of their breed. First, they were trained and working as classical geneticists, working with plants in the field and hybridizing through sexual reproduction. But in the mid-1950s, molecular genetics was born, and hybridization moved into a lab, under a microscope, eventually turning into the genetic modifications we hear so much about today. Second, the push toward profitability (above all else) slowly reduced the funding for public research in plant breeding. As many of them retired or died, they simply weren't replaced, but their vegetable varieties are still the best for home gardeners.

Breeders like Dr. Ogle were stewards—scientists in public service, commanding complex research programs with two equally important audiences: farmers and homeowners, agricultural industry and backyard gardeners. Old-style seed breeders served both. From Purdue, Santa Cruz, and Beltsville, Maryland, they developed seeds that became the food we eat everyday: yellow corn, white corn, pickling cucumbers, slicing cucumbers, green peas, wax beans, watermelons, carrots, beets, and, of course, tomatoes. Almost all came from the public agricultural colleges of each state.

The 1970 Plant Variety Protection Act moved vegetable seed breeding almost completely into the private sector. According to Dr. Ogle, "This was a mistake. It really hurt our public seed breeding, and things have been downhill ever since." Here's one simple example: I know of a collection of cantaloupe seeds that one of these breeders amassed over the years. It represents decades of collecting the best genetic potential that could be used to breed resistance to leaf diseases into the vines. After this breeder retired, no one picked up the project, and eventually, all the seeds were destroyed. Those seeds, his work, and the potential for a humidity-tolerant, nutritious melon that could be grown both in your backyard and on commercial melon farms are gone. Since home gardeners represent a small market for seeds, the dual mission of public seed breeding programs has been lost. Today, melons are bred only for the West Coast and for Texas and Florida winter production, simply because that's where the money is.

Here's one more short example a friend sites. The faculty at University of New Hampshire bred the best watermelon for the Northeast, 'New Hampshire Midget'. Since it matures quickly in a season, home gardeners in cool climates can grow watermelons. As my friend says, "It will be a cold day in hell (or a hot day in Concord) before a small state like New Hampshire pays a career faculty to do something like that again."

However, for farmers, Dr. Ogle stresses that the seed companies can and still do great work: "The science is still there, the private companies do wonderful, complex work to create diversity and new crops." But that major difference means you have to work harder than ever to get great seeds: "They [private companies] develop strictly for the big farm growers," Dr. Ogle says. "To my knowledge, there are no improvements being made for homeowners' specific needs—maybe a handful of very small companies."

When I ask him how much this has changed since his time, he says,

> Even politicians insisted we [breed for homeowners]! We had
> demonstration days for home gardeners and financial support
> for breeding vegetables adapted to home and farm. The first
> time we put on a home demonstration day, gosh, we must have
> had 2000 people who came to see the differences in 'Big Boy',
> 'Better Boy', and 'Whopper' tomato and all the different, much
> improved southern pea varieties.

Today "the companies just aren't interested at all," he says. They have
other priorities. Think of it like this: if PBS were to lose all funding, then
profit-oriented companies would be our only choice in news. If the federal
government didn't set efficiency standards, we'd still be in cars that got 6
miles to the gallon. Public good seldom motivates industry. To illustrate
how much things have changed in the seed breeding industry, during our
conversations, Dr. Ogle asked me about our family's "seedsman" in Beech
Island, where I grew up. Just hearing that simple question and realizing how
much that concept has vanished over the years was telling.

I still think back fondly on going to Simpkins' Seed Co. with my daddy.
The floor was made up of unpainted and worn boards with just dirt, hay,
and rat pellets in between. Men waited around to carry the heavy bags, surrounded by a hundred cats, baby chicks, and rabbits, all while cigar smoke
and the pink dust of fungicide that coated corn seeds wafted through the air.
Across the road was a then modern "variety store," Sky City, and nearby was
the Art Deco S & S Cafeteria. But Simpkins' was all 1950s—in some ways,
maybe even 1850s. It was mostly feed and fertilizer, but there was a bar in one
corner with a glass counter with scales and little paper seed sacks. Behind
it there were shelves stacked with big glass jars full of seeds for the season.
Once he helped you decide what you wanted to plant, Mr. Simpkins poured
seeds onto the glass bar. With his thick fingernails, he raked lines through
a spill of seeds of, say, 'Black Seeded Simpson' lettuce, 'Clemson Spineless'
okra, 'Charleston Gray' watermelon, and 'Kentucky Wonder' bean, then put
them in crisply folded brown paper bags—tiny ones for backyard growers,
bigger ones for farmers.

Mr. Simpkins helped you figure out which variety would perform
best *for you*, which one stored longest, canned easiest, or suited the taste of
your family. These days, this might all sound old fashioned and backward,

Politics, Profit, Produce

In the 1950s and 60s, Dr. Ogle and his colleagues were breeding great vegetables via public programs, but in the late 60s, Coker Seed of South Carolina led the push for national legislation to place seed breeding in the hands of private companies. The leaders of private industry were not interested in collaborating with public breeders; they wanted to run the show. Through their political influence, public programs were being cut, and the final blow, the Plant Variety Protection Act of 1970, was soon to come. Until that time, only unique plants that were propagated and increased by cuttings, or non-sexual reproduction, could be patented, owned, and, thus, easily profited from. In other words, sex was considered unpatentable, part of the natural world, the public good. You could manipulate sexual reproduction and seek new plants, but you couldn't own them or totally control them. The Plant Variety Protection Act granted patent-like protection to sexually reproduced plants, and it resulted in a split between seed breeding for public good and seed breeding for profit.

but I sure do miss him being nearby—and you probably do, too, even if you're not aware of it. Now I have to go pretty far out of my way, but I'll drive deep into the country to find a good seedsperson. If you live in a progressive, hip town, you might even run into an updated version of the old-fashioned feed and seed, now a kind of boutique, complete with maybe a coffee bar. I've been charmed and bought great seeds in downtown Seattle, Ann Arbor, and the suburbs of Dallas.

When it comes to growing from seed, Dr. Ogle says, you have three options: "If you want a great garden, find a seedsman to help you pick the seeds. Or a specialty catalog of heirloom seeds." The third option, then, is to save your own seeds. That's what Dr. Ogle has done for years: "I save all my seed. It's easy and sure. Especially the self-pollinated peas," his favorite incredibly productive and nutritious crop. Dr. Ogle uses the same old technique my grandfather used with his red salvia: "I just pull the best-looking pea plant, the most productive, put the whole thing inside in a bag, the seeds fall off inside, and I save those seeds."

Dr. David Bradshaw, another professor emeritus from Clemson University, is the man who inspired my love of old gardening styles. He's a hippy-turned-academic who loves moss, homemade fruit wines, and being naked in the woods. Dr. Bradshaw is a force—a storyteller and a nurturer; a man who knows himself, who believes deeply that we all need to save our

Dr. David Bradshaw refers to seed saving as memory banking. Lots of people can grow this same onion, but he and his family call it "Momma Sudie's Onion," and they all think of her when they grow or eat it.

own seeds, even if only to pass on these memories to new generations. As director of the South Carolina State Botanical Garden, he built a 2-acre organic veggie research trial garden. He did this in the early 1980s in Clemson, South Carolina, an era that was about as conservative and closed-minded as you can get. Dr. Bradshaw challenged it all; he brought sustainable gardening and heritage seed saving to the people of South Carolina. I was an agriculture student at the time, with three earrings and a yellow mohawk. My classmates and I saw ourselves as rebels, and Dr. Bradshaw's gardening stance was brave, rebellious, and inspirational to hundreds of students like myself.

Dr. Bradshaw started a seed bank by asking the older folks in the foothills of the Appalachian Mountains to share their heirloom seeds. "But they all said they didn't have any!" he says. "So I'd kick around and chat and see a jar of peas and ask about those only to be told they were nothing special, just something my Grandma kept for years. So here's an 88-year-old man, telling me he has the same seed as his grandmother. But he doesn't have any heirlooms."

"Everyone can have heirlooms," says Dr. Bradshaw. "And you don't have to be old or poor or from the mountains! Sometimes, I call seed saving 'memory banking.' It's a way to keep stories of people and family events. We have an old-fashioned Egyptian Walking Onion in my family. Common old plant, lots of people have the same thing. But we call it Momma Sudie's Onion; everybody in the family knows what that is and thinks of her when we talk about it." Heirlooms offer flowers, too. Dr. Bradshaw's garden holds a crinum lily that he dug fifty years ago in North Carolina. Its striped flowers open every June. And on his fence, purple hyacinth bean, while edible, is actually grown for its grape soda–colored flowers and lacquered beans; it's a commonly shared plant.

With time—and it doesn't take long—some heirlooms have become unique. Dr. Bradshaw says that in just sixteen generations of selecting for a special characteristic, you can create a unique plant. That's, at most, just sixteen years, and only six to eight if you're working diligently.

Seed saving is especially easy with closed-pollinated plants (otherwise known as self-pollinated). These are the easiest to work with and still be sure you'll get the same thing from year to year. In other words, once you

Open or Closed

Closed-pollinated plants have tight, closed flowers that are difficult for insects or wind to get into, so the female parts of a single flower are fertilized from the male parts right inside the same flower. For this reason, they're sometimes called self-pollinated. The genetic sources are known and change little, so each generation of new plants is very similar to its parents. Basically, closed-pollinated plants produce their next generation without the help of the birds and the bees. Sex just happens inside the flower, and there's some but little variation. This is why old farmers and gardeners always grow peas, closed-pollinated plants that are almost exactly the same as those of their grandparents. If they do nothing but collect some peas and keep them as seed over winter, they'll maintain grandpa's peas. Open-pollinated plants have big, floppy, colorful flowers, like a squash or pumpkin. Bees, wind, and clumsy gardeners move the male pollen from flower to flower, so there is a lot of genetic variation. If you see bees and other insects attracted to a flower, it's probably open pollinated.

If you buy a pack of seeds and want to know if they are open pollinated, you can just look them up online. For vegetables, however, that information is usually provided on the package or in the seed catalog. Seed saving is more difficult with open-pollinated plants, as characteristics that you may like of a certain plant are apt to be lost in future generations. On the other hand, the unpredictability means that a new surprise awaits you each season. For example, if you lived in Charleston in the 1950s and grew Charleston Gray Watermelon, and then sent seeds to your brother in Indiana, who grew them for years, saving the seed each winter, he would ultimately grow a different watermelon. Soil, water patterns, insects, and climate all differ, and within a few summers, your brother wouldn't really have Charleston Gray anymore. If he sent seeds back to you in Charleston, and you grew those out, you might think, "this isn't at all what I remember"—that's called genetic drift.

select something, it's not going to change much over the generations. Billy Pop Fowlkes is an old Virginia farmer, now in his late eighties, who grew up on a small tobacco farm in the rolling hills of central Virginia, near Farmville. He recollects, "We put black-eyed pea plants in burlap bags in the barn, in winter. When you wanted to eat peas, you whack the bag and whack it, with a pole, get the peas and clean the trash out in the wind. We'd eat some. What was left was seed for next spring." His black-eyed peas, over eight to ten years, adapted to being whacked in burlap bags and *slowly* became the best variety for his farm. And they most likely saved the seeds of only the most vigorous and tasty plants. But his next comment goes a long way in showing how the culture of seeds has been changed by modernity:

"But when frozen peas came around, they seemed just as good; the years of selection got lost." The frozen peas only *seemed* just as good because you didn't have to whack them or store them in the barn. So Billy Pop and his family, like many small farmers, quit saving seeds, inadvertently losing a bit of history as well as a locally adapted pea.

Something else was lost in this transition. Before those frozen peas hit the shelves, we also selected for taste, texture, and "mouth appeal" on a very local level. Those long-saved peas of your grandparents really did taste better than peas that look the same today. As Dr. Ogle alluded to, vegetable selection used to take under consideration the home gardener's needs and likes, such as fresh taste, whereas now, private breeders select for shipping, storage, uniform harvest time, and all the things that play into getting peas from the field to the freezer efficiently. Still, with certain plants, those that are closed pollinated like peas, we can easily change them, save the seeds, and readapt them to our taste, climate, and needs.

Even vegetables and herbs such as peppers, basil, mustard, and radish will seed in if you've used old fashioned varieties and let them go to seed in the garden. Here, mustard and arugula volunteer.

This sort of seed saving became the norm for many small farmers and home vegetable growers. While we've certainly lost touch with these practices, of very slightly altering our own plants with just a little bit of care and experimentation, there is a growing renewed interest, and a new crop of seed saver and heirloom seed banks. Our National Center for Genetic Resources Preservation in Fort Collins, Colorado, is one of the largest seed facilities in the world. And there are smaller specialized organizations that maintain seeds of various interests (the seeds of Thomas Jefferson and of the Cherokee Nation, for example); there are even specialized tobacco and marijuana seed banks. It is critical for us to continue to save our own seeds and to support these groups and companies. The genetics of those old seeds really were developed with the intention of ensuring that you could have a nutritious, successful garden in your backyard. And the stories of heirloom flowers feed our souls and memories of gardeners past. The old timers took part in the process out of necessity, out of their understanding of adaptability. We need to heed their call, because our doing it helps *us* understand that better, too. It helps us develop new seeds for our new and changing climates, and it's the most basic, inexpensive, and sublime kind of gardening you can do.

Updates and Adaptations

As a gardener, I do save seeds constantly, but I'm not very organized. I save them in bags and boxes and envelopes, in drink cans, and in some really old film canisters. One drawer of the fridge, a basket on the porch, underneath the truck seat, and coat pockets are all dedicated storage places. We all have our own little ways, places, and habits for storing seeds. But I even save seeds in the ground, a technique all gardeners interested in seed saving should consider. On the farm, Tom and I encourage plants whose seeds come up in the fields year after year. We also do things in a more systematic fashion that leads to fun hybrids and to cool new crinum lilies.

One of the easiest ways to save old plants by seed is simply to garden in ways that build up a seed bank in the ground. Let things go to seed so they'll come back on their own next season. Now, this may seem like a complicated way of saying that I just do nothing, but, the truth is, the process does take some self-monitoring. But you must abstain from some of our most common, modern gardening habits.

In other chapters, we've seen how some modern gardening techniques can stop the seed-in cycle. Deadheading is one of those modern ideas that

Old-fashioned single larkspur will seed itself in year after year. Often, when you buy modern larkspur seeds, the flowers will be white, pink, and blue the first year, but as more time passes, they will slowly revert to this cobalt blue.

compromises the soil seed bank. Though it seems elementary, in order to *save* seeds, you have to let things *go* to seed. Many gardeners have an obsessive tendency to deadhead, to tidy up, pull up, and then to get on with planting the next thing for the next season—*season* as defined by gardening books and marketing companies. True garden seasons ebb and flow as much as the days and climate. Gardening is fluid, and if you can resist the urge to do a fall cleanup and fall planting, you'll find that lettuce, turnips, and all those beautiful, big-leafed purple mustards will seed in and naturalize.

I know that it's hard to resist the urge; I know lots of plants look a bit untidy if you leave them to go to seed. Larkspur is a good example: a flower everyone loves when it's cobalt blue looks pretty rough a month later as it stands dying. But you have to let it stand so the tiny, golden, cigar-like seed capsules can drop. Crimson clover, vetch, petunias, and cosmos can all be ugly, too, and crush their neighbors as they dry out. Since this drives some people crazy, I look for things with more interesting seeds like nigella, with its spidery seed capsules, or money plant, with its ivory, tambourine-like seed pods. Or I use plants that go away quickly, like toadflax or red poppies. For any seed-in plants, when the seeds open, I prefer to cut the plants, lay them over, and wait till next fall when you'll see little ferny seedlings. If the soil is healthy, those seeds will germinate when the temperature is right for that particular seed.

Another modern habit that discourages most seed germination is to layer heavy mulches everywhere. Preemerge herbicides certainly do, too. In the crinum fields, in recognition of this and to encourage various types of plants, Tom and I use different types of mulches in different areas. In a backfield, old petunias, gomphrena, cleome, gallardia, portulaca, and zinnia seed themselves in every year. There, we don't mulch heavily and don't encourage highly organic soils. Those seeds succeed on hot dry soils. In other areas, heavy mulch limits seed germination and encourages crinum growth. Sometimes, it's all beyond our control, and a crop of cosmos or some weed germinates where we least expect it.

In clients' gardens, for the same reasons, I aim for various types of mulches. Some plants germinate and seed into gravel or sand or brick.

Old-fashioned petunias need a dry space, an unirrigated garden to seed themselves in. Given that, they'll be a part of your garden for decades.

The Challenge of Difficult-to-Grow Seeds

Oak trees, camellias, and roses all grow easily and quickly from seed. For other plants, figuring out how to get them to come up, how to get them to thrive, can become a lifelong struggle—or hobby, depending on how you look at it. I still have an Australian lily that my friend Ian Simpkins started by exposing it to a chemical that mimics the effect of smoke. Sometimes, I put seeds in my green, egg-shaped smoker to make them germinate better. Seeds from fire-prone regions, such as our coastal southeastern forest, are stimulated by smoke. *Pinus serotina* is called the pond pine because it grows in glades often inundated by a few inches of water. In droughts, the soil dries and the brush catches fire and stimulates pond pine to germinate. But any source of heat will do; I discovered this while trying to simulate a brush fire in my oven. When that proved too dangerous, I stuck the seeds in the microwave, after which they germinated readily. Find your special seed and crack it. You won't always find glory; sometimes you'll end up with runts, deformity, and plain-old, boring offspring. But every now and then you'll get grand results, making all the experimenting worthwhile.

Coriander and fennel seeds will stay in the ground all summer and germinate in the fall, if in the right, gritty mulch. Some plants that rarely seed anywhere will do so in heavy leaf litter mulches. I've heard the gibberellic acids from decaying organic matter stimulate otherwise difficult-to-germinate seeds. In new gardens, I'll often do a perennial planting and then overseed for quick effect. I once planted hundreds of perennial chrysanthemums, a rainbow of colors, using bare root plants. I watered in the new plants, and then spread a very light layer of compost, followed by seeds of zinnia and four-o'clocks. Then I added another layer of compost—just a dusting, and I followed that with another watering. In this garden, the owners already had a pine straw mulch, so to help my seed-in sections integrate, I topped it all with a very light pine straw sprinkle. The first summer, the garden was full of annual color, and the first fall, it was full of mums, perennial marigolds, sweet grass, and other fall flowers that grew under and behind the annuals. The second summer, the seeds "stored" in the ground germinated and made the same display.

They won't all be successful, though. I once experienced an embarrassing failure when I promised a client a spring meadow of larkspur, but failed to ask her what had happened on the property prior to my involvement. When my first sowing failed, I looked more closely to realize she had no weeds: no chickweed, no vetch, no winter grasses. The former landscaper had applied a seriously heavy layer of preemerge herbicide, which must have had a long residual life and inhibited my seeds.

The use of irrigation systems is another way that you can inadvertently reduce germination. Water has a similar effect as mulch. Many seeds rot in moist conditions, so if you want spring and winter biennials, annuals, and old-fashioned cottage plants, tame the irrigation monster. As we discussed in chapter 4, there are some seeds that like water and will germinate with lots of irrigation, but many of our biennials, our romantic, old flowers of cottage gardens, will rot with too much water.

Nevertheless, I don't expect full, complete gardens from simply letting everything go to seed; I always supplement by sowing seeds that I've saved or bought. Let nature and your notes be the guide for when to sow. Watch for germination, and then fill in where needed. Gardening by letting plants go to seed makes for more fluid cycles of gardening than the imposed spring–summer–fall planting schedule that seems more appropriate for tilled vegetable gardens.

In our crinum farm fields, all sorts of plants seed in, but we are diligent about selecting plants that will complement rather than compete with the main crop. So we have long rows of fennel, dill, and bachelor buttons in spring and peas and peanuts in summer. They all seed themselves in. I'd never put four-o'clocks in there; it's not just that they'd take over, but also that they have big tubers that would try to grow down, into, and all around our crinum lily bulbs, which is our priority. I imagine the two twisting together, like slugs mating, deforming my perfect ovoid bulbs.

So far, every seed I've discussed is from a short-lived, quick-return annual or biennial. But there's joy in anticipation and caring for seeds that are slow to grow, too. For woody plants, there's a lot that happens beneath the ground, before you see any top growth. For example, we grow one of our most underappreciated native fruits, the pawpaw tree, which grows easily from seed. After you eat a fruit, store the seed in the fridge for ninety days. As soon as you put that seed in dirt, a root emerges that can grow half an inch each day. In eighteen days, you'll have a 10-inch root; only then does a little green sprout emerge from the soil. That strong, established root system is what gives seed-grown trees a growth advantage throughout their lives.

Years ago, I met an older woman who captured my imagination by telling me about her garden, on an island in Massachusetts, where all the trees and shrubs were grown from seed. At the time, I didn't even know they had

islands in Massachusetts, much less entire gardens grown from seed collected around the world. Today, Polly Hill Arboretum, the only public botanical garden on Martha's Vineyard, focuses on collecting seeds from centers of plant diversity. Currently, they're growing swamp azaleas from seed collected in the Southeast. Director Tim Boland says, "Polly wanted to understand plants and plant growth from the embryonic stage. She wanted to witness the lifecycles of the plants in total. She once said to me, 'Why would I ever start with a full-grown plant? You miss the teenage years, and after all, they are the most interesting years.'" You can do this, too; these plants will become beautiful markers of time and lovely reminders of events in your life.

Crinum lilies have become markers of events in my life, and I've developed a great and careful system for saving crinum seed, marking dates of each seed set, and monitoring young plants for years. Though not a woody plant, I've worked with seeds that will germinate quickly but still take five years or so to mature enough to flower. The seeds of *Crinum bulbispermum* have entertained me since the mid-1970s.

Crinums are freaks. Crinum seed feel like month-old clay and look like oversized, apple-green macadamia nuts held in a thin, veined sac. The green color should tell you right away that something's strange here—these seeds photosynthesize. Most seeds are brown and dryish, dormant storage packages. In the wild, many crinum grow in swamps or along rivers. From the minute the seeds drop out of their very masculine sac, they start soaking up the sun and producing their own energy. They might float around or get stuck in the mud immediately, but no matter what, that green skin is producing food the entire time. That amorphous seed does something else strange: wherever it lands, the seed lodges half in, half out of the ground. Remember, it needs the sunlight. The new root starts growing, and the new leaves are formed belowground—both sucking energy from the photosynthesizing seeds. The tiny, white root goes deep. In one place, at just the right depth, a decision (that we still don't understand) is made: one spot fattens, forming a new bulb the size of a pea. As that little bulb fattens, the seed itself quits producing energy. The new root, bulbs, and leaves suck the remaining life out of the seed. The new leaves finally emerge to start photosynthesizing themselves, feeding the root and building the bulb. The whole process can take a month or many months. From there, it takes four to eight years for that young bulb to flower.

This has been my lifelong seed-saving hobby. I've grown thousands and thousands of crinum this way, and of all of those, at least one—okay, maybe two—seemed like an exceptional garden plant, worthy of my naming it, and it took me a dozen years. It was one seed I germinated out of a batch of hundreds. It's in the group deceptively called Orange River lilies, not because of their color, but because they grow along the banks of the Orange River in South Africa. Distinct from others of its kind, this *Crinum bulbispermum* opens sort of khaki-green. The next day it fades to pink and by the third day to rich rose—it's beautiful even in it's decline. I've tested this seedling in varying amounts of light, in different soils, and in combinations with different plants. Perhaps I waited too long to name it, but I'm a gardener who wants only great plants, not just freaks or slight variations or plants that have pretty flowers but flop over. And I wanted this crinum lily to be named to honor my mother, so it had to be right. In the end, we decided to name this special spectrum 'Aurora Glorialis'.

With flowers and ornamental plants, this is all fun, joyful, rewarding, and practical. The stories, memory banking with seeds, and sleuthing to figure out how to save and germinate can provide you with endless pleasure in the garden. But through it all, we are crucially keeping alive and passing down the understanding of how to grow and how to hybridize seeds. There is real value in understanding how what's inside a seed relates to the plants we love to grow and what's in our food. I believe this sort of knowledge needs to be passed on—to our friends and to future generations. If you get into seed saving, if you find yourself obsessed with one freakish seed, then share that passion with your friends and family. They'll share that knowledge with theirs, and we'll keep passing it down.

Unlike most seed, crinum seed produce their own energy, photosynthesizing sunlight in their green skins.

Plant Breeding for the Home Gardener

Joseph Tychonievich wants everyone to be a plant breeder. His book, Plant
Breeding for the Home Gardener: How to Create Unique Vegetables and
Flowers, *sounds like a throwback to the 1950s, something you might find in
your uncle's flower shop. But Joseph is a young guy who spends his time doing
bees' work, moving pollen from one flower to another to combine genes of different plants. He says:*

You can create a new variety just by letting the bees do all the work for
you. This will work for many plants with large, showy flowers that attract
bees. Columbines, violas, hollyhocks, poppies, and petunias are some
of the easiest. Vegetables like kale, broccoli, corn, and squash work
well, too. Natural selection will preserve the individuals best suited to
your climate; then you pick out the ones you like best to create a variety
suited to your personal tastes.

To start breeding your own strain, choose one of those types of
plants and beg, borrow, or buy a diverse group of varieties. You can
choose varieties with complementary strengths and weaknesses, say a
very vigorous viola with a limited color range and another with colors
you like that doesn't grow as well. Or, you can just get a bunch of different things and see what happens when their genes mix up. Planting your
different varieties close together will encourage cross-pollination.

Then you just let the bees have at it. If you want, you can help out
by moving pollen from one flower to another with a paintbrush or even
your finger. For plants that self-sow reliably in your garden, you can just
let the seeds fall where they will, and the next generation will spring up
all by itself. Or collect the seeds and sow them out yourself. Remember, the more plants you grow, the more beautiful things you'll have to
choose from, so more is always better!

As the next generation grows, look them over. Save the ones that
grow the best, flower the most, and are the prettiest, and pull out the
rest. Be a little ruthless here! Saving just the prettiest, healthiest, plants
will make your next generation even better. Keep doing this year after
year, and each season the plants will become better adapted to your climate and garden, and more beautiful!

You can also always add new varieties with a new color, fragrance, or form to the mix to keep spicing things up.

Joseph knows about diversity. As nursery manager at Arrowhead Alpines Rare Plants Nursery, he manages hundreds of variations of common and rare plants, like the somewhat common but underappreciated Clematis integrifolia. When you do see it for sale, you usually see the same old, one variety. Joseph maintains and loves at least ten different varieties, which reminds us how detailed, intense, and fun horticulture used to be and should be again.

Handmade Structures

Using Garden Materials for Trellises and Sculptures

AMERICANS SEEM TO like structure and solidity in their gardens, from little vine supports to pergolas, arbors, or trellises. It's not entirely clear why this is, but I think it's because we want things solidly defined. We want to see the difference between a trellis and a bunch of sticks tied together. We like things strong and intentional, and over the years, we seem to be more and more drawn to new, industrially designed, and manufactured products to provide this structure. The appreciation of the rustic and the rickety requires a considerably more relaxed aesthetic. I tend to think our desire for structural strength might be tied to that transition from rural to urban that we've discussed in chapter 1; we've gradually transitioned from jumbled country yards to simplified suburban landscapes. We've moved from hand tools to machinery and from rooting our own cuttings to purchasing professionally raised plants. Many of us have left the cobbled-together structures of the country behind.

Remember those splayed, fan-shaped, prefab trellises? Remember seeing them at the hardware store? So enticingly trim and tidy in a shirt-tucked-in sort of way, leaning up against a wall, looking like the answer to your prayers; with that trellis in place, your morning glory would grow perfectly, spill over gently, and erupt in flower across the top of that redwood fan. Maybe they were your first step away from country ingenuity. But when I see them, I just think of more work and frustration. Because at the end of summer, I'm the one that has to untangle the vine from that dreamy fan trellis, used for a moment before the vine shot out to smother the neighbor's dogwood tree.

Instead, I make vine supports out of sticks and wires, but with an attention to detail that makes them artful. Well, I think artful, though some people have called them "high-tech redneck" right to my face. But mostly, as a plant lover, the primary reason that I use or build any garden structure is to

Downcycling

Is that trellis really made of copper, steel, or a dirty mix of who knows what? The problem with recycled wood and metal is that the process of recycling releases dioxins, which are highly toxic environmental pollutants. It's also a particularly inefficient process and results in materials of diminished quality and strength. Most of those cute, little metal chickens, pigs, and "wrought-iron" trellises were once cars, shaving-cream cans, or stoves. Plastic furniture might have been milk cartons, water bottles, or fast food containers. When we recycle, those things get melted down, which releases toxins, and made into something else of often poor quality. Recycling, then, is often more like downcycling—and it's simply slowing the movement of trash to the landfill, not really preventing it. By growing your own trellis material, you're simply not adding anything to this cycle, and you'll have a more original, beautiful, and rustic trellis as your reward.

give a vine a home. My goal is usually to make something that's suited for the size and climbing method of those enchanting life forms and their sinuous, flexible bodies. Vines enthrall me. Not just the flowers and fruits, but the adaptations that they've made to fit into a niche among other plants. Vines clamber up old, decaying stumps seeking sun. Some even make specific adaptations for climbing up other vines—transitional opportunists thriving between death and renewal. The structures I build for vines gently control them, keeping their limber bodies where I want to see them. A vine structure has to be big enough to carry the weight and small enough to keep the leaves, flowers, and fruits at a level for people to enjoy.

I get great pleasure out of weaving and making things with dry branches, vines, and old bamboo. And, again, time permitting, it can be a joy to make your structures stand out as artful, sculptural, and fun things all on their own. After all, vines will take a bit of time to grow out and cover them, so you'll have to look at that support structure for a while; why not spend a few minutes to make your trellis beautiful? And sometimes it doesn't even take more than a change in your perspective and expectations as to what makes something beautiful. You can take three sticks, quickly tie them into a wooden teepee with wire, and have yourself a simple bean trellis. But from another angle, those three sticks, smoothed by years of use, tied together with coarse twine, can make a rustically beautiful thing. Try looking at the common resources in your backyard through someone else's eyes. Imagine you're walking through a little village in Thailand and watching an old woman put together a three-legged bean trellis made of bleached knotty sticks. It might just be a simple bean trellis to her, but you'll see the simple beauty of it—those sticks could be sculpture on your living room wall, a token of your journey.

Garden stores and catalogs will sell and market their structures and trellises as "prefabricated," which often means collapsible and shippable. But, as a result, it also often means they're undersized for many vines. And their production and transportation can hugely contribute to environmental problems of waste and air pollution. Even those "composite" wood boards and little metal trellises made of cheap recycled metals are not nearly as innocent as they make themselves out to be.

Luckily, I don't often find myself in places that sell those fans and little trellises, so I'm not tempted—and I have an abundance of yard waste to work with. Those in condos, or with neat, tidy gardens, don't always have the same luxury, which is another compelling reason to think about materials for construction and weaving garden structures when you select your trees and shrubs. You can also always keep your eye out for your neighbor's yard waste piles, too; once you've had your eyes open to the possibilities, you'll be amazed by the beautiful branches that they throw away.

From a design perspective, handmade structures become visual tributes to the people who made them. From an environmental perspective, it is essential that we make the most out of the things we have in our yards and nearby in order to reduce the environmental demands of shipping materials around the globe. And practically speaking, what we're often seeking in our yards and gardens—shade and support—can be accomplished with just a simple stick in the ground. Home-sourced, handmade structures offer all of that, and they're quick, easy, honest, cheap, customizable, and charming.

The Teachers

SUE AND BAN VAN

In 1974, while I was climbing magnolia trees and catching frogs, two boys about my age spent their time in Jardin des Tuileries in central Paris. My memories of that age are idyllic; theirs are bitter. It was the year of the Cambodian revolution. Since Cambodia had been a French colony, these two boys and their mother, Sue, like many Cambodian refugees, initially fled to Paris. Sue remembers the time: "Only me and my boys escaped. I walked around in a fog. We couldn't believe what was happening. We didn't know what to do, so I looked at public gardens, just walking and sitting on benches, and I guess, absorbing French garden styles." But in disbelief, eviscerated, and living at the top of a city tower, she couldn't garden. "I tried to plant onions on the balcony, but when the water runs through, drips below, French people screamed! They got very mad!"

Today, Sue is in control of her 1-acre South Carolina garden—well, *control* in a Buddhist sort of way. When I ask Sue to point out differences between her garden and the neighbors', her answer is emblematic of the way she lives her life, but it's not the answer I expected: her hands flutter over her heart as she says, "When I stay here, I take great care of the garden. I need to walk around my garden. To water, to pull weeds. But when I leave it, I don't worry about it. Things die, weeds come, vines go everywhere. It's on its own." Acceptance of that kind is a recurring theme of this book—in fact, Sue's life and her garden are representative of so many themes in the book. The differences in her garden that were most obvious to me—certain statues, certain Cambodian vegetables—are nominal. The huge difference is in her attitude, her approach. Sue is content with the reality that everything we do, everything we build—including our gardens and garden structures—is temporary.

Sue is a thrifty gardener. She grows and preserves lots of vegetables and saves her seeds. She starts things in a tiny kit greenhouse and props things up with sticks. She has shade structures and trellises made from scrap wood. Vines cover fences, and depressions catch rainwater. It's my kind of garden. If I hadn't known better, I'd have guessed she was a farmer's daughter.

Her life before she came to South Carolina wasn't something I'd ever heard Sue mention. On a recent visit, a friend looking at her pictures pointed out an old black-and-white photo and whispered, "That's Sue with [famous architect] I. M. Pei, and that's her with the King of Cambodia." Seeing our amazement, she started to tell us about what came before—before she was American; before her frequent visits to Paris; before she ever touched dirt, seeds, or any plants other than a magnolia flower floating in a silver bowl.

Until she was twenty-three, Sue never even set her eyes on a kitchen, a garden, or even the driver's compartment of her family's cars. There were other people to do those things. Her incredibly smart family led a privileged life. Her father was a developer; her brother was comptroller general of Cambodia; he basically ran the country until the communist revolution.

After that gut-wrenching history of the revolution, after she fled to Paris, she ended up in Augusta, Georgia, where there was a small group of

refugee officers, including one of her brothers. Sue took a job in a Chinese restaurant, where it occurred to her that if she wanted her two boys to know more about their Cambodian food and culture, she was going to have to learn to cook. And to cook Cambodian food in Georgia, in the early 1970s, meant she was going to have to grow the food herself; she was going to have to learn how to garden. She'd have to garden to eat, to survive, to teach, and to reconnect with the land that she loved, but which was now suddenly and finally gone out of her reach.

Cambodian dishes—like the curried, flavorful fish amok—are layered with complex mixtures of many herbs, roots, and vegetables that you just can't find in the local grocery store. Like the gardens of many immigrants, Sue's grew slowly as her network with other refugees grew. Refugees had more important things to think about than fleeing with herbs or seeds, so meeting someone who could share a piece of spicy galanga root was like running into an old friend. Even today, she looks forward to the Cambodian New Years celebration in Washington, DC, where nurserymen and home

Sue Ban and my mother have taught each other about their gardens and cultures.

gardeners set up booths offering plants, spices, and specialty herbs for cooking. Since Cambodia has opened its doors again, more and more tastes of home come to market. Sue and a network of gardener-cooks trade seeds and plant parts by mail, too.

I've discovered in my travels that asking to see a flower garden is one thing: it's planted to show off and share. But asking to see the vegetable garden is different: it's more personal. It can be like the difference between asking to see someone's living room versus their bedroom. Like many immigrants, Sue doesn't share her garden readily. But my mother, through her sincerity and gentleness, befriended Sue decades ago in a small southern town. Their similar personalities and shared love for gardening, cooking, and children led to a lifelong friendship between two women from drastically different economic, linguistic, cultural, and gardening pasts. Though Sue has since moved to the beach, she and my mother still trek across South Carolina a few times a year

Sue's garden structures, mostly made from found wood, put bitter melon, cucumbers, and other hanging vegetables at the right height for easy picking.

Trellis, Arbor, or Pergola?

These three words seem interchangeable today, but I use them to refer to different scales of structures, with trellis the smallest, arbor in between, and pergola the largest. Knowing the origins of the words is helpful in remembering this. *Trellis*, from the Old French *treille* (vine trellis), refers to coarsely woven cloth, so it includes things woven from metal with a lattice-like pattern. A trellis is used to train a plant vertically. *Arbor*, from the Old English *herbier*, refers to a shady nook. An arbor connects overhead, creating a small entry or seating area. *Pergola* comes from the Italian variety of grape used to make the wine Pergola Rosso. These large, overhead structures held the vines, making grape picking easier. A pergola is often part of the architecture of a house, providing large shaded areas.

to go on garden tours, to trade plants, or to teach each other how to work with a new vegetable they've discovered. Not too long ago, Sue brought us some wing beans, and walking around Gloria's garden, she spotted a giant gourd vine smothering a plum tree. Thirty minutes later, Sue and my mother—who's only ever used a gourd to make a birdhouse—called Tom and me in to taste their cooked gourd, sautéed with garlic and egg. I envy the friendship between these two women, and I'm amazed at the way they meld their cultures. Something we used to make into dippers and martin houses, now suddenly shows up in our soul food suppers as sautéed tiny gourds and crowder peas over cornbread. And what's more, gourds grow better than zucchini and yellow squash in our hot climate.

Obviously, making use of what you already have is another recurring theme in this book. For your own health, the health of your community, and the health of the planet, learn to appreciate, to value, to drink, and to eat the things that grow best in your climate. If any plant isn't serving the purpose you want it to, it's time to question your thinking—perhaps you're going about it all wrong. As we walked through her garden a few years ago, I pointed out a papaya tree and expressed frustration that my papayas would never fully ripen before frost got them. "Oh, then don't want them to ripen," she said, "Learn to make green papaya salad!" It's since become a fall staple.

Just as they taught me to use gourds and papayas, Sue and her husband Ban also opened my eyes and helped me to better appreciate those pulled-together structures I'd known my whole life. Through a different lens, I began to see the beauty of handmade, temporary elements in our gardens. I never knew how to manage gourd vines properly until seeing them on upright, overhead structures in their garden. Table-like arbors, 7 feet tall, have corner posts connected with slats. They are covered with a 4-inch square of metal mesh. As the vine grows, the sagging of the mesh under the vines' weight is checked by bamboo sticks pushing upward. The leaves create a shady workspace and the fruits hang at an easy height for the gardener to pick.

Sue and Ban have thrived here, in an American dream sort of way. Sue worked constantly, eventually owning a restaurant and a salon, while Ban,

Neither a pergola nor an arbor, this simple structure is part of my garden gym; it's a chin-up bar and a trellis for *Rosa chinensis* 'Speedy Gonzales'. It's made from simple supplies from a hardware store and has been in my small city garden for twenty years. Underneath, you'll find a mix of vegetables, flowers, and a sustainable lawn.

Spirit houses connect the spirits of all the people who have ever lived on a piece of land to the current residents and garden.

an engineer, retired from Kimberly Clark and now shrimps for fun and extra income. They can afford to build trellises and structures from their choice of material. There are places, such as a swing by the pond, which they built with purchased 6-inch-square cypress posts. Sturdy swing supports and a pergola top those, but to the sides, for shade and flowers, a lighter, more rustic trellis holds a tangerine cross vine and recalls that fuzzy aesthetic.

I love the fluidity and contrasts of Sue and Ban's garden: flowers and food, play and work, companionship and solace, excitement and worship. They mix old and new, and they consider *how* things grow and how they can best be used. Their handmade structures reflect that—no one-size-fits-all fan trellises here. Big gourds, bitter melon, beans, or trumpet flower vines grow to different sizes; the place you'll pick from (or just lean in to smell) is different for each. Your handmade structures can be built to fit, to show off, and to exert just a little control.

Sue's husband, Ban, does a lot of the infrastructure work in the garden for her: the lath vine supports, the ponds, the bermed rows in the vegetable garden, and the deer fence. But he has a shed and work area to call his own—his part of the yard, which is where you can find him a lot of the time since he retired from the paper mill. Except for a slight accent, he's basically just one of the fellas. In his shed, PVC pipes made into sheaths hold fishing rods; there's a boat, tiller, and a little tractor. He's adopted the local fellas lingo, too, like when he talks about his and Sue's working relationship: "She says what she wants, and I do what she wants." He throws in phrases from country songs and blue-collar comedy shows, but he's trim, elegant, and youthful—a former Cambodian soldier, and a man of engaging contrast.

Ban has one structure that the local boys don't have. At first glance, it looks like a birdhouse on a pole. But it has a little door and a tiny front porch, on which sits a metal tube holding incense sticks and a tiny bowl of cut flowers. Sue seems a little uncomfortable even discussing it: "That's

Spirit Houses

Throughout Southeast Asia, in prominent corners of yards and gardens, you'll see an upright post about head high, topped with a tiny model house. Spirit houses work to protect spirits; they're a place of connection and recognition of the earth and the journey of the land and ancestors. Inside the spirit house is often a small photo of a grandmother or grandfather, the spirit that's lived on and protected the land. Sometimes figurines representing family or servants are placed inside, too. In the early morning, daily offerings of food, flowers, and incense are made on the house's little porch. It's ultimately a place for good spirits, but it's also placed in the yard as a distraction, a way to capture the attention of mischievous spirits and keep them from going into the family home. Spirit houses are a link to the time before Southeast Asia was predominantly Buddhist, when animism was widely practiced. In animism, people worship the unseen forces, the soul of the earth, and spirits living inside rivers, trees, and rocks. They affect the material world—if the river is dry, crops suffer, and thus people suffer. Though it's called by different names, every country in the world has a prehistoric legacy of animism. But as religions blend old traditions, it's been obscured in most places. In Southeast Asia, however, even as Buddhism, Hinduism, and Christianity moved in, the spirit house remains a strong signal. Another friend of mine in Siem Reap, a young and hip guy, added a spirit house to his newly built, sleek, modernist home of white concrete slabs and warm tropical woods. And his spirit house reflects that architectural style. This strong tradition remains, even among the youth, as an obvious and lovely symbol of the past and future life force of the earth.

his way," she says. It's a spirit house; a place for guardian spirits to live. Ban's is made from construction excess, red pressed roofing, and bright blue plywood—pre-Buddhist reverence rendered by a southern country-boy engineer. I think it's a beautiful thing to have in every garden—a reminder, a place for the spirits of the earth, the place, and the family.

Sue and Ban melded all of these garden styles on their own. Their garden is not Cambodian or Carolinian, but something new. It honors the lessons and loves of those that have come before them. Sitting in her living room, Sue emphasizes this repeatedly during our conversation, each and every time fanning her heart—sort of a pulling of heartstrings, an automatic amplification of her words: "I have to grow plants; I love to plant; it's part of my therapy. When I start to observe, there is nothing else."

Getting into your garden in this way can be transporting, meditative. And creatively working with structures can add to this. You'll begin to think more deeply about where your construction materials come from, and you'll more carefully consider the lifecycle, the climbing habits, and the size and body of your vines. And your handcrafted structure will amaze with its rustic beauty.

——————— Updates and Adaptations ———————

I'm very fond of temporary trellises for the simple reason that I've long been infatuated with vines. I've grown many more vines than I've had walls or pergolas. One summer, I put rebar stakes in the yard and kept the vines pruned on those poles. I love vines of all kinds and seek ways to use them in gardens.

Some plants stick to walls without any help from trellises or wires. In this photo, clinging roots of *Campsis radicans* stick to a painted brick wall on an abandoned building.

If you're interested in vertical gardening, vines can be sublime. Vertical gardening systems, while very trendy, can come across as very unsustainable and artificial; getting dirt, water, pots, and misting systems to hang on a wall just seems like way too much work. But vines can do all of that for you. My favorite vines to grow on walls, like Virginia creeper and other *Parthenocissus*, produce modified stems and leaves as they grow that turn into pads with the ability to cling to even smooth surfaces. Some even produce drops of glue-like substances that will further adhere them to walls. Keep in mind that they're mostly harmless to your structures or walls, but like any plant they can get out of check. They'll need care and monitoring.

Other vines scramble, climb by rootlets, or twine themselves around things. Beans, cucumbers, black-eyed Susans, or butterfly peas (*Clitoria ternatea*) need to scramble up and over something. Gourds, bitter melon, winged beans, Cambodian soup vine, and New Zealand spinach will twine up things and eventually flop over without more support. You *can* buy little pyramidal or cone-shaped wire structures called tuteurs that stand in the garden or in a container, but, invariably, they are too small. In the South, most store-bought tuteurs are quickly covered, and the poor vines will telescope until they finally find a nearby birdhouse to scramble over.

As with the fan trellises, tuteurs in most garden stores are made overseas with "downcycled" metals, and shipped using packing supplies and gas. Why contribute to the pollution and waste? On our farm, we use a lot of bamboo for teepees and tripods. Built with dry bamboo, a tripod for peas can last four years before it starts to rot at the ground. Sticks or rebar work great, as well. The most simple and earth-friendly tuteurs will be made from your garden waste and work for a season or two before they become a carbon source for mushrooms and microbes to break down into compost. When I'm gardening for other people, I find lots of the material to make these on roadsides, or I use prunings from someone else's yard. For a little more permanence, a thicker trunk works well, too.

And if your vines overtake your tuteurs, you can trim them. Vines are particularly good for providing extra green material for chop-and-drop mulch or green starter for the compost pile. Sometimes I intentionally plant a large, fast-growing vine on a trellis that is too small, knowing I'll machete off the extra to use for compost throughout the summer. Sometimes, I just goof up and plant the wrong thing on the wrong structure, in which case, I do the same thing.

Seasonal structures can be really big and fun. For giant growing vines, Will Hooker, a landscape architect and professor at North Carolina State University, goes all out with modern versions of natural material structures. Will and his students work with bamboo. One summer they made a giant spiral structure at the J. C. Raulston Arboretum. It was 40 feet tall, and as you

Will makes weaving, painting, and playing with bamboo meditative. Leading a group of students, he created this giant dragon, all from bamboo, for the J. C. Raulston Arboretum in North Carolina.

walked into it, it got ever tighter and smaller until finally you reached the center. Moon vine and hyacinth bean cloaked it, making the center room dark even on a bright autumn day.

How else could you get that height and mass in just one summer? In a new garden or open space, to create privacy, a forced view, or to get shade or cloaking, temporary structures are the way to go. When we built Riverbanks Botanical Garden in South Carolina, it was completely enclosed in 10-foot-tall brick walls. Since we didn't want the plantings in the foreground to look like they were growing in a giant brick box, our first order of planting business was to cover the walls. Vines were the answer, but most fast growers like coral vine, jicama, and woolly morning glory all need something to climb. Little strings and wire are not enough. A simple solution was lean-to trellises. Basically, you stick some tall saplings or bamboo in the ground, 2 feet or so from a wall, lean them in, and let the aggressive annual vines do their work. Over a new wall, this adds softness, a sense of age and maturation. I used a 15-foot-tall lean-to, planted with loofah vine, at my house to add shade and privacy over a western-facing wall of windows.

By simply weaving the same saplings together, you can make a fence to stop foot traffic or dogs, or help weighty grasses and big perennials stand up. You can also use them to add a bit of focus and structure to a perennial border. Sometimes a perennial garden simply needs a little bit of human touch to make the picture complete.

A handmade structure can tell a story of the people who made it. These young Laotian monks are learning to use hand tools and to make structures from bamboo.

It's this human touch that most intrigues me with handmade structures. It's fascinating to see the craftsmanship in a garden structure, and it's satisfying to be able to relate to an unknown person who has built something that fills a need that we share. It's another "aha!" moment to see how a gardener has tied her cucumber vine up to a cedar teepee. While we might never have thought of it before, it suddenly makes so much sense. That moment ties us in, and it tells us a story of the tools that a person had to work with, and the planning or lack thereof that went into the building of their trellis. A handmade structure might even tell us, upon seeing it, how much of hurry she was in, or what kind of day that gardener was having. Handmade structures, elegant or rickety, connect us with a shared ingenuity, to nature and to other people.

A tuteur made from bamboo spray painted red guides fast-growing flowering vine at Riverbanks Botanical Garden, South Carolina.

The Art of Building With Bamboo

Will Hooker, professor at North Carolina State University, was an inspiration to me years before I ever met him. A fellow bamboo enthusiast, Will's bamboo sculptures and structures are often designed and built in collaboration with his students. I long wished to be in one of his classes on design, permaculture, or sculpture, but I never had the opportunity. The following is a hands-on project that Will suggests for helping newbies learn a few of the intricacies of the bamboo arts. Will says:

Bamboo is arguably the most useful plant in the world. It is edible, it can be used to make tools, fishing rods, pipes, and clothes, and it is especially useful in building, as it is light and stronger than steel by weight and able to be crafted into any-

thing imaginable with simple tools. Asian and equatorial cultures have maximized its use for millennia. For those who are concerned about potential invasiveness, it is an easy-to-purchase, eco-friendly option for use as: garden stakes, fences, trellises, and building houses. A Belgian bamboo fence [a wide, open lattice in a diamond pattern] is both decorative and functional in fencing off spaces and serving as a growing support. I've used mine to espalier apple trees.

Will's bamboo working tools.

I've used mine to espalier apple trees. Will concludes, if you have a stand of bamboo, you already have most of the material needed to build a fence, trellis, or sculpture. It is best to use three-year-old bamboo culms (poles). These are typically a lighter, more olive green color than first-year culms, will usually have aerial roots at the detritus layer, and, for the best indicator, will have a mottled "camouflage" look, which comes from bacteria on the culm. Using a pull saw, cut the culms low to the ground. To trim off branches, also use the pull saw.

Will's Belgian fence made of bamboo just after planting.

If you don't have bamboo in your yard, it is readily available from nurseries and supply companies (like www.bamboosupply.net). Also, take a look around your neighborhood for houses with a thick stand growing in their yard. Ask those neighbors if they'd like to share their bamboo; chances are that they have more than they know what to do with. The culms can be used whole for poles, beams, or rafters, or they can be split (lengthwise) to make more flexible strips for more imaginative forms. Bamboo splitters, hatchets, and saws are available online.

There are many means for connecting bamboo pieces together. Traditionally, twine of some sort was used, and this will work well with temporary trellises. I have also used 17- and 19-gauge electric fence wire, as it is made

with an alloy that doesn't rust. The quickest method is to use plastic zip ties, which can be found in most electrical sections of stores. These are recyclable, so carefully collect the trimmings and the ties themselves when taking down a structure.

A few years later with apples espaliered.

Any leftover bamboo is also great for starting fires, as the epidural layer is loaded with paraffin wax. If you're adventurous, you can also throw parts of culms, ones that are closed from node to node, onto an outdoor fire to create fireworks. Some say that the name came from this practice when the culms exploded—Bam! Boom!

So unfetter your imaginations and let the fun begin!

Tools

Reviving Essential and Neglected Hand Tools

WHAT BECAME OF MACHETES? Once a very common household tool, people tend to laugh—the editor of this book, a gardener himself, laughed—when I talk about how machetes need resurrecting. Actually, he laughed when I said that I needed money to go to Haiti to further study machete skills, but he also admitted to not knowing how to use a machete. Hoes, mattocks, hand weeders, pocketknives, or dibbles elicit the same bemusement. How many people actually use any of these tools anymore? The golden age of hand tools peaked around 1930 and has been on a slow decline ever since. That has to change. Hand tools need advocates. They need us to keep them from falling into obscurity and rusting away buried in the mulch pile. We need them because they're ingenious; they keep us in shape and connected to the ground.

Can't you just feel it in your forearms—the satisfaction, the vibration of a hook blade knife, curling through the thatch, the lawn, into gritty dirt

below? The satisfaction when you hear the little snap of tension when a dandelion root gives way to that sure blade? Can't you relish the reward of a pile of weeds cut, cleaned, and taken care of, your gardening tasks done for the day? It's just muscle and drive out there shaping our world. Or, rather, it used to be. Good tools and the skills to use them get forgotten when souped-up mowers or colorful bottles of chemicals seem so seductive. But like knives, forks, and spoons, simple garden tools have been perfected by the generations who depended on them. Solid, perfected, and sexy, hand tools work.

For centuries, tools were made of wooden handles and metal implements. They were made at home or by local blacksmiths. They were made to fit individual needs; at the height of the Victorian gardening craze, for example, gardeners demanded hundreds of variations on hand hoes, cultivators, and sheers. But, basically, all hand tools are elaborations or modifications of spoons, forks, and knives. They were modified depending on your climate,

For safety and comfort, it's important to use tools that fit your body. The Raintree family comes to visit and volunteer on our farm so frequently that we keep shovels made to fit each one of them.

My friend Alfredo helped me hone my machete skills on his family coffee farm not too far from the border between Haiti and Dominican Republic.

LEFT Alfredo has worn the handle off of his machete and replaced it with a washcloth and duct tape.

crop, or gardening styles. For example, three-tined pitchforks were designed for moving light straw on midwestern farms, while five-tined forks work for longer hay in the South. Before factories, a good idea spread slowly, traveling by word of mouth and example from blacksmith to blacksmith. The early 1900s, with the Industrial Revolution, brought uniformity and the availability of hand tools. We had new factories that made variations on tools to custom fit every weeding need, every size row, and even every person's height and weight. With industrialization came mass production and suddenly a great idea from some old Victorian greenhouse gardener was available in stores around the world. The best blades from Caribbean sugarcane fields, tools never seen elsewhere, could now be shipped and shared. Hence, we had fleets of things like wheel hoes—remember wheel hoes? Those lightweight, wooden-handled, push plows with a big metal wheel up front and interchangeable blades at your feet? My father used one until the 1980s and, in fact, there are a few New England companies still making them. We had customizable rakes and Amish digging forks in every hardware store in the United States. Most importantly, we had diverse farms and gardens full of people who would actually get out there and put those things in the ground. Even the words for different kinds of tools became more uniform. Think of lots of little village blacksmiths making some combination of shovel and axe. One might have called it a mattock, one a handpick, and one an adze, but eventually we all started to call those things hoes.

But the changes in tools in the early 1900s paled in comparison to the seismic changes brought on by World War II, which introduced and left us with the small motor. This changed everything—even how we pick up leaves. Before the war, there was indeed small motor technology, and a typical, prosperous farm of the time might have had one single engine. But it was a big, immobile thing. The motor stayed in one place and the farmer brought different attachments to it—for example, a corn sheller, butter churn, saw blade, or a pump. But during World War II, for the settling, moving, feeding, cleaning, and draining land for thousands of men in the Pacific, the government needed the industry to better develop and to downsize the motors. As a result of this pressure, motors not only improved, but after the war, we were left with small motor factories that needed a shift in purpose. And millions of people who'd fought in the war and seen these things save time, human energy, and lives came home to make their world better. My grandfather was the first man in Hampton County, South Carolina, to have motorized equipment, and he was so in demand, he never even had time to farm his own land. Farmers who still had hoes and mules hired him and his motors to work their fields. But this was never going to be enough to sustain these motor-making factories, and the country was gradually leaving the farm, anyway. What else can we make? How about putting motors on those old blades and shovels and selling lawn mowers, edgers, or tillers to the new suburbanites? Briggs & Stratton, for one, developed the first lightweight aluminum engines, and the suburban gardening revolution began.

We've all seen how eager new homeowners have embraced these seductive new technologies. I think it's their seeming complexity that drew us in. Hand tools began to feel like antiquated accessories. Motorization made the new things seem more professional, too, more awesome and sophisticated. So we forgot

The perfect accessory. This well-made tool has been passed down through generations. It started life as a full-size hoe that belonged to a friend's grandmother. Over the years, it's changed into a hand hoe imbibed with memories.

the old farm tools, and they soon turned into termite dust and rusty dull hooks. All the while, people who could carve hickory handles and sharpen machetes left us, too. Without those skills, hand tools got stashed in the barn with the old butter churn. We motorized machetes, swing blades, trenching and edging shovels, clippers, brooms, and, most absurdly of all, wheelbarrows.

What didn't get a motorized replacement got a pump and a magic tonic. Sprayers took the place of weeding knives and even tarps—yes, tarps. You can actually buy liquid glue that you mix up in a hose-end sprayer and then squirt on piles of leaves to glue them all together so the wind doesn't mess up your pile. It's basically a tarp in a bottle. Is it just me or does that make you want to lift one of those hair-sprayed leaf piles and put it on your head like a giant, multicolored, maple leaf, beehive headdress?

We sure love to spray things—I'm sure there's a text somewhere on the evolutionary reasons that spraying seems so satisfying, so fun. The idea of liquid tools is so compelling that there are miles of racks of them in stores; it's a huge industry. We can't take our eyes off bottles long enough to find that hook blade knife and gouge seven weeds from the dirt or squish three worms on the cabbage. There's even a hand-held garden sprayer with a rechargeable battery-powered pump attached. How did we get here?

The truth is, I'd like to be able to say that hand tools are simple, but they do require skill and training. Maybe that's another reason for their demise: we've simply forgotten how to use them. They look so simple, a wooden handle with a metal blade. But wood and metal can hurt. I have a knee scar from machete training and plenty of dents on my fingers from newly sharpened shears. Sometimes we can't even figure out what some old tools were used for—we've drifted that far. Hand tools have their complications. But they better connect us to the ground, to craftsmanship, and, today, to a generation of ingenious people who worked hard and appreciated elegantly effective tools and the satisfaction of honed skills.

———————— The Teachers ————————

GARY AND JENNIE WHYNAUCHT AND BOB DENMAN

There's something about midwesterners and their fascination with old tools. There's probably an ethnography of them and their tool fetish on some bookshelf, too. Though lots of my old mentors still use hand tools to garden, the teachers in this chapter represent more: a history with and an updated love for old hand tools. Gary and Jennie Whynaucht and Bob

Denman are part of the diaspora of the Midwest who share and teach their love of tools.

Gary and Jennie Whynaucht are midwesterners who settled in North Augusta, South Carolina. They and my parents met and shared their love for old things, for soup, history, and tools. Gary is a Michigan man with grit and a quiet respect for your personal space and privacy that wears off in minutes, and soon he's telling you every detail of the past, present, and future uses of this amazing sprinkler that he found at the 400-mile yard sale in Kentucky. He left Michigan to be a Navy machinist and settled in a little town near the Savannah River with his wife, Jennie, who's from Illinois and Wisconsin.

The wide river keeps things humid here. Spanish moss drapes a crepe myrtle tree. Old tables, a woven tobacco drying rack, and a canoe hang on the barn, which looks like an old tractor shed. Being there, you want to explore, to scavenge, and to touch rusting treasures, and you can. Gary and Jennie run a unique hybrid of junkyard and antique shop. Step into that little barn and get lost among the junk; then if you make it out, there's a metal building, Gary's one-man factory, with piles of old shovels, model planes, and sprinklers. This serves as the supply line for the charming antique store they run out of their little Victorian cottage.

Gary Whynaucht with a mysterious, hefty rake made from old plow parts and the hinge from a barn door.

Gary fixes stuff, including old garden tools, and Jennie sells stuff, though Gary is quite the salesman himself: "I sell those old sprinklers for people to put on their patios just like misters you see now at amusement parks. I tell people, just hook them up and turn them on—let the water droplets cool the air. You'd be amazed at what I can sell." Gary and Jennie not only sell from their shop but, literally, all over the country, setting up booths at garden shows, tractor pulls, and sometimes antique shows. They're always looking into other shops for objects, parts, and ideas, so they know antique farm and garden tools from all over the country.

When you meet Gary and Jennie, you know they've built this whole place themselves. Gary's a fix-it guy, an engineer. You can see the layers of his interest: gardening tools, piled on sprinklers, piled on industrial gears and chains, all on top of a couple of old Volkswagen Beetle engine blocks. He can fix those, too.

They first met my parents while driving around looking at old places, old barns, and meeting country people. Jennie reminds me, "We loved being with your parents. We'd get together on Sunday nights for a bowl of soup. We always took something we'd stumbled onto, and Gus would explain southernisms." They all loved to snoop around the countryside, then sit around the kitchen table and marvel at the ways people adapt and change.

In plundering his shop recently, I found some of Gary's meticulously dismantled, cleaned, oiled, and renovated pitchforks. Some are being restored to their original condition. He also has lots of homemade hybrids—things people made from parts of other things. He shows me his renovation process: how to release old rusty parts with a little heat from a torch, how to wire-brush them without taking off all the original color, and how to shine them up with a little lacquer. He was working on a cultivator rake with removable 10-inch, curved, hard tines, each attached with bolts. The bolts allow you to adjust the rake to fit between different size rows or to replace a broken tine; this one is being returned to its original condition.

Antique sprinklers represent the taste of different eras.

This shop is like a salvage yard. Its inventory of broken handles, blades, rake parts, and boxes of old sprinklers wait to be joined by the next new finds. When Gary finds something special at a yard sale, something missing a part, he knows just where to look in his own shop for that part. He can renovate or update from his shop. Since lots of their customers are buying sentiment, memories, and romance—even art—Gary amends for that market. Sometimes, he makes things pretty: he'll take a broken handle from a shovel, made of beautiful wood, with a burnt-on insignia, and he'll hone it down into a dibble. Or a hoe, with edges beyond repair, will become a whimsical heart-shaped hoe, perfect to display on your porch or mounted by the front door as a coat rack.

Gary picks up a curved, hard-toothed rake made from old plow parts and a hinge from an old barn door. The thing weighs about 15 pounds. We marvel at the adaptation and the muscular customer who requested it. It makes us want to know more, to contemplate the need and the story that brought this crazy, heavy, rigged-up thing into being. We're both curious

Sprinkler Art

You could trace the history of popular design, art, and architecture through lawn sprinklers. Whether they're cast-iron alligators and tractors from the 1920s and 30s, or sleek lines, aluminum accents, and molded Bakelite curves from the 1950s and 60s, sprinklers have become collectables. Early sprinklers were mostly for agriculture and the occasional wealthy homeowners, but just before and after World War II, two things happened that made the sprinkler industry thrive. First, in 1937, the Fair Labor Standards Act set the workweek to forty hours, which meant that most people for the next few decades had their Saturdays off to spend doing yardwork. And second, after the war, the federal government set up an assistance program under the GI Housing Bill that allowed hundreds of thousands of returning soldiers to buy homes—all with front yards. The lawn business was born, and sprinklers became everyday items. Gary and Jennie have little water shows in their backyard; the water dances and twirls. Some sprinklers move themselves across the lawn. Some have heart-shaped blades. Gary has specialized in fixing these for actual use in the garden, for cooling patios, or as amazing little water-fountain displays.

Sprinklers in action make for artful watering and cooling tools.

about the rake's history, the lost creation story of this behemoth. I'm curious about how hand tools differ by region. Do you find more homemade, rigged things in places that used to be poor? Did termites or weather or labor issues influence the styles of tools in Minnesota versus Texas? Where are people today most into collecting and using old hand tools? Though that's a question that Jennie and Gary can quite confidently answer: "the Midwest."

They speculate that in the Midwest, the climate dictates that tools get put up at the end of the summer. Often, tools, wheelbarrows, and such were put into the granary, literally covered with grain, then not used until the next spring as the grain reserves drop to reveal the tools. In the South, the climate means that tools are left out all winter, subject to rot and wear. Other possibilities include the cultural history of farming in different regions. Midwestern small farmers tended to do everything themselves, so they knew,

My favorite hand hoe, made by Bob in his Red Pig Tools shop, gets used constantly. Joel Raintree and his son Reid use it to make divots in hard clay for planting small grasses.

cared for, and stored their own tools. Easterners, however, tended to have access to more temporary labor, so people there may not have put as much effort into tool care.

Some of this history is just fun speculation. We like to imagine stories of people who might have needed, used, and adapted certain tools. Bob Denman, on the other hand, wants the facts. Bob is a former journalist who set out on a project to compile a complete history of hand tools. Bob ultimately became a blacksmith in Oregon, an authority on the history of old tools and a seller of more hand tools than you can shake a hoe handle at. Bob makes new tools, with old methods. I use his tools in my work, and I treasure them.

One of the first things Bob said to me from his home in Boring, Oregon, was, "I lived in South Carolina for two months. Camp Jackson just before I shipped out for Vietnam." I've heard this before from veterans; they always go on to say how South Carolina was hotter than Asia, and they haven't been back to either place since. I thank them for their service and tell them that South Carolina is still the hottest place in the South. Bob might have escaped the heat of the South, but now he's a blacksmith, just about one of the hottest jobs a person can choose. Bob's nighttime hobby became a gardener's treasure chest: his business, Red Pig Tools, produces about 250 different tools, and he knows more about the history, manufacturing, and uses of hand tools than anybody else in the country.

He wasn't always a tool guy, but there was an old man in his life who wanted to pass on the craft of blacksmithing. That mentor saw in Bob a spark of creativity and an interest in working in front of a forge fire. So with a full-time job, Bob smithed at night. He says the basic skills were easy, but the art, the certain twists, curves, and angles that make a tool perfect, took years of practice. And while the tools as objects occupied his hands, the tools as history occupied his mind to the point of obsession. And his training as a journalist meant that he knew how to follow a good story and research minutia. He's still perfecting his craft—he's also doing his part to pass it down, as well. While we were talking, Bob's new apprentice, a young man who shares that same spark, came in with a still-hot Swedish-style brush axe; Bob is blown away by the artfulness of the work. Then he launches into a history of Swedish-style brush axes and their replaceable blades. The history of hand tools makes for good storytelling, and those stories are getting passed on more and more as modern gardeners seek quality tools.

Bob can launch into the history of just about any tool you put in front of him. And he can explain what to look for in a good axe, hoe, hand claw, shovel, or asparagus-harvesting knife. He'll start with a seemingly random story, like how the half-moon edger was originally made by Roman soldiers to cut turf to build walls and ramparts. Through stories, he explains the history of metals and how the best wooden handles still come from Appalachia. Bob tells me that in the heyday of hand tools, you could go into

What Makes a Quality Tool

When seeking out a high-quality tool, it is important to examine and consider these three aspects:

- **Metal.** Mild, forged steel is best. It's hard enough to be sharpened and hold an edge, but not so hard as to be brittle. Cheap, soft metal can sometimes be bent or nicked simply by pressing it against a concrete floor. Metal parts should be forged, made as one piece, or if there are attachments, one piece should penetrate the other.
- **Wood.** Look for a "Made in the USA" sticker. Great handles still come from hickory and ash forest surrounding the Appalachian Mountains. Good wood will have a dense grain, like an old baseball bat. One simple trick for learning what brittle wood looks like is to find a cheap broom and examine the handle. Many cheap broom handles are made from the tropical tree, ramin. It's fine for brooms and towel racks, but it breaks under stress—and the tree is considered endangered. If a tool's handle looks like a broom handle, keep looking.
- **Metal and wood connections.** Take a look at how the metal blade is attached to the curved metal part. If you see a little glob of molten metal, that means the blade is "tacked" on and will easily break off. The collar that holds the blade to the wood is called the ferrule. Its purpose is to protect the handle from cracking. Flimsy plastic or metal won't do the trick. Look up into the ferrule. A strong connection is a squared metal blade "stem" pushed into a round hole in the wooden handle. The best tools have a small, ratchet-like collar adding strength, but you can rarely see it. Look for a rivet or small rod punched into the handle, piercing the ferrule and the wood.

any hardware store and find a broad range of tools to fit your exact need and even your body size. There were more head shapes and many shapes came in different widths. Each width was available in several handle lengths. In Bob's shop, this is all still the case; he often customizes tools by modifying blade width, height, or pitch and by modifying handle length.

Bob can put into words preferences I've developed for tools but never really thought about. He explains the reasons behind some of my tool preferences. For example, as a guy with decent biceps, I like a heavy tool. The heft adds cutting and digging strength when I work. I know some people prefer lightweight aluminum for shovels and hand tools. I ask Bob if he works with any materials besides steel and wood. But no, he stresses, the word *blacksmith* specifically refers to the black oxides that come from iron or steel. "Anyway," he explains, "aluminum is too soft to hold an edge." It makes a fine shovel or hand claw if you have good dirt, but I often work in rocky, rooty soil, so a sharp steel spade is my preference. Since I'm always breaking tools, I want tools that can be repaired, and neither aluminum nor fiberglass can be fixed. Inevitably, my tools will be run over or snapped off when I try to use a shovel to pry up a 300-pound crinum clump. I can still hear my daddy yelling at me, "Boy that shovel is not a pry bar. If you need a pry bar, go get one. Don't break my good shovel." Alas, old habits die hard, and I still use shovels to pry, so I need shovels that can be repaired.

Another trait of my tools that Bob asked me to consider is their weight and balance, especially when choosing a mattock or any hand tool that might put stress on my wrist. A mattock with a hollow metal handle is just too light to actually move dirt. It bounces if it hits a stone, stressing my wrist. After his questions, I realize that most of my tools are the same as his: wood and steel. In fact, many of Bob's tools are replicas and adaptations honed by professional gardeners. He finds old tools in garden shops and books. Some were even mass manufactured at one time. So he's following the lead of tools perfected by gardeners over centuries.

The author and his niece, Caroline, with donkeys Buck and Justina, who keep the pastures mown and provide manure for compost. But they're not the only tools shown here: red and yellow clover in the pasture work to produce nitrogen.

He also considers all sizes of people and even keeps abreast of ergonomic research. When I asked him for advice about my hand pruners, he told me that while mine were quite good, he recommended another brand because they've modified the tilt of the blades according to military studies on hand movement. Taking his recommendation, I now use Bahco ERGO pruners and have less cracking in my wrist after a long pruning session.

Choosing high-quality and thoroughly tested hand tools is essential for people like me, who use them a lot. It makes your work faster, easier, and just better. I work with a woman who hates my shovels; she swears by her lady's ergo shovel. Bob reminds me that tool quality and choice can seriously affect your body. He's great at editing tools and helping people find just the right tool for the job and for their body. The most important step in choosing tools is finding tools that make the work smooth and safe for *you*. As Bob said earlier, find the right shovel, rake, or hoe height; find the right balance and weight for your body; and keep things sharp and clean.

Gary and Jennie and Bob represent the romance and the practicalities of choosing and using tools. For me, and for you, those are important choices. Tools can connect us to history and old memories and keep our bodies in shape for another day's gardening.

——— Updates and Adaptations ———

My tool chest, full of old wooden and metal tools, looks pretty old fashioned next to the mobile mechanic shops of many other landscape professionals. I can find some of my tools in kitchen stores, while others are bought on trips to places where people still make tools by hand. However, I still use the occasional motor, and there are plenty of plastic handles, sprayers, pumps, and a sleek German knife or two. One of my favorite modern tools is not a variation on knives, forks, or spoons—it's more like a crescent wrench. I even have living tools: yes, Buck and Justina, dwarf donkeys, mow the pastures, saving us hundreds of dollars a year in diesel for the tractor and bush hog. And if you consider that one of the overarching tenets of this book is that gardening itself is caring for the earth, then microbes, worms, bees, and plants are tools, too.

In our landscaping work and in our crinum nursery, Tom and I have very few motorized tools. We keep our blades filed, oiled, and handsome— or at least we try. We run a gardening business and a farm in a climate where there is no long resting season. So our tools wear out fast and repairs are often a low priority considering everything else we have to do. Nonetheless, it works for our nursery and garden design business, and it works for our home garden. It really is possible; you really can garden without an air-polluting, cough-inducing, gas-powered blower or mower.

One little blade that I wouldn't be without is my hook blade pocketknife. It's nothing complicated, just a single blade in a plastic case, but the hook is the key. With it, I can get a weed from between perennial crowns, divide a grass into twenty little bits to spread around, and weed a lawn in a way that is satisfyingly addictive. I use a blade made of hard Swiss metal. When I'm using it in the grit surrounding little bulbs, it dulls quickly—I've sharpened these down, over years of use, to thin vestiges of their former selves. They're cheap enough that when lost, given away, or sharpened into obscurity, I just get a new one.

But the blade that draws the most questioning glances when I use it is the machete. Its uses are almost endless: cutting stakes, shearing hedges, chopping mulch and green stuff for the compost, limbing trees and brush, clearing a path through an exuberant perennial border, leveling overgrown cannas, shearing pansies, splitting bamboo, and building upper body strength and stretching. You can even do a little light mowing with a machete, leveling annoying bahia seed heads or dandelion flowers that pop up overnight. When I first started really watching people use machetes, on a tiny coffee farm in the cloud forest of the Dominican Republic, I quickly started to wonder why we didn't use them back home.

One evening, sitting around without power but with plenty of rum and thoughts (the kind of thing you do in remote villages where nights are long and dreamy), I remembered that my daddy knew how to use one, and I tried to unravel how he might have learned. Could daddy have learned in Cuba, where he was a US Navy Seabee doing ground-floor construction

These hedges are well kept with a combination of gas-powered sheers and hand pruning. My bench design for this garden was inspired by another tool: lobster traps seen in Maine.

for Guantanamo Bay? Or maybe he picked it up in Spain or Africa? I tried to recall his stories. Then, in writing this all down, it dawned on me that he grew up making molasses from sugarcane and a mule-driven screw press. The process depends on cleaning leaves from the cane, which was all done with machetes. I've seen boys in the tropics doing this, wearing nothing but shorts. As luck would have it, back home I was able to dig out some old pictures of daddy and his grandmother by the molasses press, blades cleaned, cane waiting to get squeezed. Later, Gary Whynaucht showed me the folding pocket machetes that lots of American soldiers learned to use. In other words, Americans *did* know how to use machetes. Not only did most of us not only forgot the skills over the years, but we forgot that the tool was even a part of our parents' toolbox.

For shearing, however, machines do work well, and some of my favorite motorized tools are my electric and gas hedge shears. I'm not into tight, crew-cut hedges; they're just too much work in the South. But for my tall bamboo hedge, for perennials and annuals, and of course for giant expanses, motorized sheers do the trick. People tend to freak out a little when they see me do this, but I love to shear pansies and perennial borders. It's not a control thing—well, it's not too much of a control thing—it's just an

Machete Training Tips

The key to using a machete is to create tension on the plant to be cut. In fact, a good machete user will often have his machete stick—just find the right one and hone it. Using a little branch as a handle and a hook on the end, it's used to grab a plant and pull it to the side, creating tension on the stems to be cut. This is all done with your weaker hand; with your dominant hand, make a long swing, twisting your wrist and aiming for the bent, tense part of the tree. Another trick is to keep a loose grip on the machete, which allows you to use the power of the weight of the blade in the swing. Some machetes come with a lanyard, a wristband to keep the thing from flying away and hurting someone else. I've bought all styles, all over the world, but for work, a plain bush machete with a 14-inch blade is great. It's easier to learn to use them by watching, so check out some great videos on the web from www.machetespecialists.com.

understanding of plant growth and responses in our climate. Pansies stretch here; when we plant in the fall, it's often too warm for them, so they stretch and look floppy. To remedy this, I shear them so they'll be tight in the winter. If I have only a few, hand pruners do the work. But if I have flats to plant, the electric hedge shears or a very fast, sure swipe of a machete behead the little, floppy plants cleanly. After this, they'll produce more side shoots, and hence more flowers. This was a lesson that Ruth Knopf taught me: for the most bountiful, colorful, carpet of spring pansy flowers, sacrifice the piddly winter flowers and pinch until March, which is easy in a small garden. In a mass, a quick shearing produces waves of spring flowers.

In our crinum nursery, we have two mechanized blades: a tractor with a bush hog, for mowing fields, and an old DR mower that we call the push hog. Both are for brush cutting, for work that could only be done otherwise by an entire crew of people. Both are over twenty years old—quality matters. The DR mower does allow me to grow things like *Vinca major* and *Indocalymus*, aggressive groundcovers that need a severe cut back every few years. Yes, a sling blade *could* do the same thing, but as with hedges, we make the call that the labor needed for that is more than we can afford, and as long as we keep their usage to once a year, the environmental cost of the smelly mowers won't be too heavy.

I planted and kept this formal perennial border for ten years. With our long, hot summers in South Carolina, we can sheer many perennials a few times a year.

As I mentioned earlier, we also rely on hook blade knives (sometimes called grafting knives). For dividing crinum, the power of water, to wash away soil, and a good hook blade do the trick. I love working with knives; I think back fondly on learning to sharpen them with my father with whetstones and files—kind of like learning to shave, a mystery shared between father and child. We also use a large, underappreciated tool called a Weed Wrench (available from www.weedwrench.com), which is kind of like a giant pair of pliers on the end of a balanced lever. It is used to pull up tree seedlings and saplings by uprooting them. As crinum are a long-term crop, most of ours stay in the ground three years before they are ready to sell; we get pecan, hackberry, mock orange, and other tree seedlings as weeds. The lever grabs a seedling at ground level and when you lean back on it, the seedling will pop out of the ground, from around the mass of perennials, with every little bit of its roots intact.

Our crinum fields are edged by pecan trees, and, in some places, even underneath them. Crinum thrives in light shade, but as with any shade garden, you have to keep the canopy thinned. So we have two tools for that: loppers on a pole and a fantastically sharp pruning saw. The loppers can be sharpened, but a good saw has a replaceable blade. I buy mine from a local dealer, because I know they'll always have the right replacement parts. I once splurged and bought myself a Japanese hand saw, only to realize that our saws cut when you pull back on them while Japanese saws cut when you push (a little secret of effective sawing is to understand that most saws only cut on the pull back stoke; the forward stroke is only to get the blade back in place. Knowing that saves lots of energy and wear on your forearm). Make sure you know what you are getting.

We also have an array of carts and wagons. As with any tool, quality counts. We still use an American-made wooden garden cart that's about thirty years old. The plywood has been replaced twice. For all our carts and wheelbarrows, we use only solid rubber tires so as to eliminate the frustration of flat tires and having to change the tubes.

ABOVE Tom with a slender sling blade. Sling blades, swing blades, and scythes need to be incorporated into all modern tool sheds.

BELOW There is nothing as rewarding as getting an entire, sprawling chickweed out with one simple slice through the stem. My favorite tool for this is a simple and inexpensive hook blade (or grafting) knife.

My machete collection stored on the side of the crinum shed.

Almost all of our farm tools go on the occasional road trip to my clients' homes—I rarely feel the need to add new things to my array. Many clients do have their own mowers and other tools like that, and usually they have other people who do that work for them. I just do the gardening. People tend to get a kick out of some of the simple farm tools that I cart over from our farm, which work great for suburban-size landscapes. One of these is a chest-mount spreader that I use to spread cover crop seeds or fertilizers. Mine allows for even applications and even directional spreading. I've had the same one for fifteen years. Another simple tool that people like to see in action is our compost tea maker, which I described in chapter 2. I start the drum one day, turn on the aerators, and use the tea the next day to inoculate soil with microorganisms. The simple submersible pump brews the tea; then turned around, it becomes the applicator.

All the tools I use connect me through history and biology to caring for the soil, to memories of people who used them before me, who shared and passed on these simple, effective, common-sense tools and skills. It's time for us to do our part, too. Whether you are a gardener or just hiring one, help bring back the romance and the exercise and the attention to detail that comes from using good old muscle-powered hand tools.

LEFT Three of our favorite, well-worn carts: a double-wheeled wheelbarrow with solid rubber tires (right), an American-made garden cart that you can repair with plywood (left), and a wagon-like cart made by my daddy, who used the wheels from his father's cart to build it.

RIGHT My Weed Wrench is great for uprooting weed tree seedlings and saplings. This is the smallest size you can get.

Making the Most of Technology

There are a few modern devices that I rely on in my gardening work. I always carry a small digital recorder in my pocket. I use it to record notes on tasks that need doing, plants that need harvesting or sowing, and things in flower or bothered by a pest or in need of cutting back. I've tried the recorder on my phone, but it's too complex; a simple recorder works best for me and can even be plugged into my computer to download and email notes.

And since all those notes need a place to go, I turn to a computer program called FileMaker. My friend and mentor J. C. Raulston introduced me to this program over twenty years ago, and I've relied on it ever since. Easy to customize and easy to share, this true database stores decades of names, notes, photos, and articles. I've built simple, journal-like databases for many of my clients so that they can better manage their plants. And in every botanical garden I've worked on since J. C.'s intervention, I've insisted on the implementation of customized databases to fit the project's needs.

My partner in all of this today is Hunter Desportes, who helps me make sure that the databases better fit each client's needs. During his work, Hunter asks key questions about how each database will be used,

In this multilevel garden, hand tools are a must, as most mechanized tools simply don't fit. The owner of this garden also keeps his work list, pictures, and notes in a database.

and how the separate fields and elements relate. In our botanical garden work, we chose this system to manage huge plant collections because it's powerful and user friendly. It helps develop plant management systems to document, inventory, and produce work reports, labels, and publicity for the plant collection. That's never an easy task for me, but getting it right from the start means that all those notes, photos, and flowering dates can be combined into a large and useful record. Today, I can stand in my crinum field, in the dead of winter, all plants dormant, and look at the database on my phone to know where any one plant is and dig a custom order. For a garden client, we can look through the database to produce a quick report, with photos and plant names of garden work that needs to be done each week. A recorder, a camera, and a database are like the quill, sketch, and ledger of modern gardens.

Scavenging

Unearthing Great Plants from Many Places

IN A MELODIC, South Carolina, Lowcountry brogue, Bennett Baxely ticks off the plants his mother used to collect: "Mother would pull fetterbush from the swamps. She contended that you could pleasantly shrub a yard with flowering things from the woods. Sarvisberry, pearlbush, chokeberry, and Black Thomas grape, son of the muscadine; it is the best for eating and pies." He sings, "She nevah boughta' thing."

He could be talking about himself or me or a million others who are so inspired by the woods that we just want to be a part of them, to have a part of them always growing right out the back door. It wasn't all too long ago that nursery centers were quite rare and most people knew how to start their own plants from seed, cuttings, or divisions. There were no garden centers because there weren't yet any plastic pots for cash and carry. Sure, we've always had nurseries, but they were mostly field nurseries where plants were grown in the ground and shipped bare root. And while famous places such

as Pomeria or Fruitland's Nurseries offered an astounding 1300 different plants thirty years before the Civil War, most nurseries focused on fruits, nuts, and bare root trees.

The rapidity with which this has all changed since then has been astounding. Buying plants is now an established part of our consumer culture. Supply-chain retail stores and their glitzy marketing have even, for the most part, usurped nurseries. With so many plants now available at our fingertips, and marketed in gas station displays and grocery stores, it's tough to remember the time when this wasn't the case. Granted, I've sure enjoyed easily available plants. As a nursery owner, I make money from plants that I start and grow, as well as plants that I buy and sell, and as a garden curator and designer, I've bought from and supported nurseries all over the world.

But as a gardener, a thrifty man, and someone who loves to trade a good story, I've been seeking out and sharing free plants for most of my life.

What plants wait to be resurrected in cemeteries? As plants go in and out of fashion in the nurseries and gardens, cemetery plantings tend to stay the same. So besides being intriguing places to explore, old cemeteries are repositories for old-timey plants.

When I was a boy, one of my favorite treasure troves to plunder was the dump outside of our local cemetery. While my parents were busy in the cemetery tending to our family plots, I scouted for potted chrysanthemums and Jerusalem cherries, and old bulbs, that had made their way from the graves to the dump, to take home. I loved the old metal wreath holders, too, but my parents wouldn't allow me to take them. I think perhaps they thought they were too morbid. After the dump, I also loved to scour the woods, along with abandoned home sites and old gardens. I still have a giant wild azalea that I dug up while we were out fishing. I didn't even realize until I was in college that we had two world-renowned nurseries nearby. Plants grown by my friends, mentors, and peers, and collected during hikes and on vacations—I still make visits to local cemeteries whenever I'm traveling—make up most of my scrapbook garden. Scavenging plants and finding plants for free or cheap can quickly become an obsessive hobby.

It can become a profession, too. Obsessed visionaries kick around the world in forests and hard-to-get-to villages; without them, we wouldn't have such things as disease-resistant wheat, which was introduced to us by Russian botanist Nikolai Vavilov. An American counterpart, David Fairchild, the first USDA director, scavenged the world for plants, bringing back and selecting new varieties on which were later built the soybean, pistachio, and avocado industries, among others. Nor would we have many of our spectacular garden plants. But it was self-taught gardeners who brought us Christmas poinsettias and summer gardenias and the white-flowering American wisteria. More and more gardeners are going back into the woods looking for plants and new forms of old plants.

While any of us can do this, too, it is important to have a solid understanding of what you'll be able to make grow, or whether or not you can find someone who can. Scavenging is not hoarding; scavengers need to know their needs, limits, and propagation techniques. I've included this chapter later in the book for exactly those reasons. It takes a true gardener to make a great scavenger; we keep dirt, plants, and stories alive to pass on to others.

The Teacher

BENNETT BAXELY

Bennett Baxely once told me, "I never even went into a nursery until the late 1950s, when someone opened one in Summerville," about seventy miles away from his home. Fifty years earlier, Bennett's parents started their home, farm, and gardens near Hemingway, South Carolina.

In the spring of 1920, Bennett's parents made their little homestead on a modest piece of unamended land along Hog Crawl Creek, which is near Black Mingo Creek—let's just say it's way out in the lowland swamps of South Carolina. They relied on their land to produce almost all the food needed to feed the family and the farmhands. As times changed and food could more easily be bought, they expanded their flower gardens. Walking me through his garden, Bennett recalls:

> The pasture originally came up to right here, 20 acres of horses
> and cows right in front of the house. Daddy planted his tobacco
> beds out here. But when he'd go away on hunting trips, mother
> would inch the fence back, imperceptibly, away from the house
> until, finally, that's where she wanted the garden to be. So the
> garden eventually moved to this line.

Over the following decades, shrubs and other garden plants took over everything that used to be their farm. Almost none of the ever-growing assortment of shrubs was bought. Today, a low brick wall marks that fence line that his mother moved. It's still a transition from the front lawn to a slightly more wild azalea garden. The garden that now completely surrounds the little farmhouse is so old-school cool, so layered, that *Garden Design*

Bennett with his pass-along roses. The pale rose on the left has been in his family since 1900, and he says, "not even Ruth Knopf knows what it really is."

magazine gave it their Golden Trowel Award. It should tell you something that a modern magazine, with page after page advertising the newest and hippest in gardening, would still find room to honor a garden of scavenged plants and porch rockers with twisted cornhusk seats.

One of the many beautiful plants that you'll find growing there is *Achimenes*. While its common name is up for debate—be it widows tears, hot water plant, or magic flower—Bennett insists on the scientific name; that's how he learned it when he first met this plant. He's kept his going on his farm and garden, where he's always lived, except for a brief stint as principal of the Daufuskie Island School, made famous in Pat Conroy's *The Water is Wide*. There he met Mrs. Fripp, a 100-year-old local gardener, who gave him a few of the wormy little rhizomes of *Achimenes*. Now, every summer, he has buckets of pink, purple, and white trumpets overflowing in pots or spilling out of Kimberly ferns. They rarely need water. In late fall, every year since they were passed on to him, he sifts through pots to find the tiny rhizomes. He puts them in envelopes, a few of which are set aside for his garden next year and a few to be passed along as gifts with the story of Mrs. Fripp from Bluffton.

Bennett knows I enjoy his circular stories. He'll start out talking about some vine he pulled up from the swamp, then move the story seamlessly to taking a freighter bound for Tangiers and sleeping in a churchyard in Kenya and ending up a decade later with a visitor dropping by who's connected to all of those places and the very same vine. I can listen to his stories and be in his garden endlessly—they are intimately connected. So I'm intrigued on a summer walk when he says, "Come and see! The cartwheel lilies are in bloom!" I scratch my head; Bennett knows I love lilies and am especially intrigued by the ones that grow in the flooded cypress swamps around here. From black water that looks thick with mosquito larvae, snakes, and alligators, delicate white petals, 6 inches across, emerge. Below, the big bulbs hold themselves in cypress knees and muck. It's mostly only collectors that get to see these bog plants, of the genus *Crinum* or *Hymenocallis*, that the locals call swamp bells or spider lilies. But in all my experience, I've never heard anyone use the term *cartwheel lily*, so when we round the corner toward a clump of old swamp bells, I call him on it: "I've never, ever heard you call them that." With a sly grin on his face knowing he got

Bennett got his first *Achemines*, planted here in front of a building that was once a commissary for farm workers, in 1960 from Mrs. Fripp, a 100-year-old gardener in Bluffton, South Carolina.

Wisteria macrostachya 'Clara Mack' is a sort of small-growing, unique, white cultivar of the Carolina wisteria. It's path into cultivation includes scavenging by many people of various races and classes.

me, he replies, "I've never called them that before, but don't they look like someone doing a cartwheel?"

Bennett, his parents, and my parents lived in a world where pulling some swamp lilies out of the ground to take home was perfectly acceptable. When I was little, it seemed like every adult I knew carried a little folding shovel in the trunk. Lots and lots of gardens were shrubbed with wild plants dug up off a ditch bank. More than just scavengers, these gardeners found, propagated, and introduced some of the horticulture world's best selections of plants. In flood lands along southeastern rivers, our American wisteria is spectacular in spring, and it fixes nitrogen all year long. The white form, *Wisteria macrostachya* 'Clara Mack', is a favorite of mine, and it came from another woman who bush-whacked in southern swamps in the 1950s, saving and making several important selections of local plants. The wisteria is said to have been growing in the ruins of a once grand plantation. A yard worker there recognized it as unique and brought a piece to Miss Mac, who grew it for decades. In the 1980s, botanist Mike Creel determined that it was a rare white form of the somewhat elusive blue-flowered *W. macrostachya*. He scavenged a piece of it and introduced it to horticultural trade.

In the 1940s and 50s, new and improved cars and roads got people deeper into the wilds, into places where great plants waited to be discovered. Mary Henry, a Pennsylvanian, is another who embodies the tradition of the self-taught, adventurous plant collectors of the 1950s. Mrs. Henry made long, strategic plant-scouting road trips from Pennsylvania through the South. By her upbringing, she should have remained in Philadelphia ballrooms, but, as she once said (as quoted in the horticultural magazine *Arnoldia* in 1950):

> Rare and beautiful plants can be found in places that are difficult of access.... Often one has to shove one's self through or wriggle under briars, with awkward results to clothing.... Wading, usu-ally barelegged, through countless rattlesnake-infested swamps adds immensely to the interest of the day's work.

While collecting in Coweta County, Georgia, in 1954, she found a richly colored sweetspire (*Itea virginica*) shrub. She took that little plant on a journey through lots of hands and decades, and back north, and the Henry Foundation for Botanical Research distributed it. In the 1980s, that plant took an amazing journey through the hands of a series of famed horticulturists; Michael Dirr, of the University of Georgia, saw it at Swarthmore College and suggested to Judy Zuk, the director of the Swarthmore's arboretum, that it be named and introduced. Woodlanders Nursery of South Carolina took the challenge, naming it initially as 'Select Form' and then as 'Henry's Garnet' in its 1986–87 catalog. 'Henry's Garnet' went on to become a poster child for the fledgling native plant movement.

Had Bennett and Mrs. Henry met, I'm sure they would have struck up a lifelong friendship, exchanging letters and plants for decades. It's almost a surprise that they didn't; if you ask around the area for the best person to talk to about local plants, everyone would point you to Bennett. You could then experience the same slow and leisurely drive to his garden that I enjoyed. After driving for miles on a two-lane paved road, sprawling fields on all sides, past Scott's, a gas station and barbecue joint where they cut meat on retrofitted, stainless-steel surgery carts, you'll turn off the paved road, on down the long, sandy lane. Tall pines and scrubby underbrush surround you on each side and, ahead, a white, dipped-in-the-middle picket gate and weathered rocking chairs call out to you from the farmhouse porch. The gate is a kind of joke, a garden folly. Just as you drive up, the dirt lane cuts abruptly left, going by the barn and into the backyard. There is no front door or front gate formality here; it's comfortable and calm. We round into the backyard to find a few ladies leaving, bidding goodbye to Bennett, their tour guide and host. It's a funny place to see a tour group.

Seeing these visitors, I'm immediately reminded of my grandmothers. Thin, blue-haired ladies, full of delight, probably saying something like, "What a *gra-and* day!" Their goodbyes and thank-yous are repeated, extended. Everyone is standing under a low-branched pecan tree, where the garden tour ended—and just where it had started. Here, where visitors arrive and depart, simple, low ivy beds get all the attention. The beds are almost mod: glossy, black-green ivy mounds kept in a cleanly repeating line, in stark contrast to the flat, apple-green centipede lawn. But the zigzag ivy beds mark a very important historic definition: the line between what used to be the yard and what used to be the vegetable garden. The jagged pattern was left when the 100-year-old split-rail fence finally rotted away. Ivy then took over, and Bennett pruned it. That's the magic of Bennett and his approach to the garden: deep layers and swirling stories, all charmingly simple.

The older shrubs and trees around Bennett's garden were scavenged from the woods by his mother, while he picked up many of the perennials and container plants on his travels.

Buicks crank and creep down the lane. Bennett is ready for us, his next visitors, though he is still waving goodbye with his white hair wisping in a breeze, his blue oxford a little damp. But he's ready to do it all again, to tell us stories of each and every plant, not just the wild ones brought from the woods. While admiring an azalea in his yard, Bennett and I realize that we share a connection through that plant. He tells me about his mother embarrassing him as a boy, on a tour of the South Carolina State House grounds, back in the 1950s. She carried a little knife, and as they walked the grounds, she'd slip cuttings from the bushes into her purse, appalling him. He reprimanded her after the tour, "What if you got caught? That was stealing!" Listening to that story, I realize that this azalea is a cutting of a plant my great uncle planted when he was horticulturist at the State House in the 1940s through early 60s. He was known as a man who would break off pieces of plants for visitors to take home and root. I know he would smile to hear that I ran into some of his plants yet thriving half a century later. And the image of Mrs. Jesse James Baxely, Bennett's mother, sliding cuttings into her old Sears purse makes me smile, too.

Updates and Adaptations

Most people don't ride around today with a trunk full of shovels. Our hobbies have changed. It's not even that common anymore to see kids running off to spend all day in the woods. But I think the same urges are still there. Whether you're from the country, the city, or the suburbs, you can still learn to scavenge, even if it sometimes takes different forms. You'll find opportunities in the woods, behind grocery stores, or in empty lots in Philadelphia. The drive is the same: to kick through poison ivy, climb cliffs, wade in mud and ticks, find cool plants and be with them in the wild. In doing so, we find each other and often ourselves.

There are other cultural shifts that have also made plant collecting less common, for better or worse. The days of stomping through someone else's swamp and digging up their plants are over. Property rights, conservation issues, and plant protection laws—or at least the enforcement of them—in a more populated world have changed. Now you *have* to ask for permission. And if you're going to get something from the woods, go for seed or only a cutting. Collecting from the wild now comes with a professional rulebook

Plant Collector Codes of Conduct

The Convention on International Trade of Endangered Species of Wild Fauna and Flora (CITES), along with a wide array of local plant-based groups, offer professional codes of conduct for plant collection. Most are aimed at professional collectors, but there are also guidelines that the rest of us can follow. The New England Wild Flower Society offers a commonsense manual on plant collecting. It runs fourteen pages, but the main takeaways are: act with respect to people and the earth; be sure that you have permission; and know what you are collecting. When you know it's a rare plant, leave it alone. If you're sure that it's a common plant, observe the following guidelines:

Plants. Collection of a few plants for personal use will probably not cause damage. But don't dig up something that will only die in the process. And never collect a plant when you see fewer than six individuals in the area. As a general rule, no more than 2 percent of the plants at any site should be collected. For example, if there are a hundred plants, you can only take one or two.

Seed. To maintain the reproductive potential of a local population, generally no more than 10 percent of the total seed production should be collected from any population within a single year. For species that produce only one or two capsules per plant, leave them alone.

Cuttings. Know whether the species is likely to propagate successfully before taking cuttings from wild plants. Be aware of the proper time to take cuttings.

Foraging. Know the regenerative capabilities of the species before collection. For example, creasy greens, popular wild plants for cooking, make millions of seeds. So there is likely an abundance of seed stored in the ground. However, ramps, a popular edible bulb, can take five years to mature. You eat the bulb; you kill the plant.

After collecting, let the plants recover to their original state. If there is more than one person collecting at the same site (whether you know it or not), the original state will be different for each collector. If you're unsure whether anyone has been there before, or if there is evidence suggesting that they have, it's best to avoid collecting anything at all. Keep in mind that for some species—for example, some ferns, Orchidaceae and Liliaceae in particular—ten years might be required for the population to recover fully.

Cordgrass and yaupon holly along the lagoon at Hunting Island, South Carolina. Both plants are used extensively in gardens and have been used as edibles, crafts, and pass-along plants.

'Regina's Disco Lounge' is a found crinum with a pale pink stripe.

addressing ethics and conservation. There's actually a voluntary International Code of Conduct for Plant Collectors, and a slew of organizations offer their own in extended, formalized, contractual language based on one simple message: be respectful. A respect for wild plants is crucial; it's important to realize that every day we're impacting the natural world in ways that we'd never imagined. It wasn't even that long ago that we never would have considered that our actions might cause the extinction of entire species of plants or animals. Now we need to know better.

Scavenging from cultivated and waste places is a little different; it involves respect for privacy and ownership. *Shack botany* is the term used for scavenging from abandoned cultivated places. There are many formalized groups you can join up with to do it. Working within a group can be fun and help you find great sites and meet like-minded people. Rose rustlers like Ruth Knoph spend their Saturdays scouting farmsteads, cemeteries, and ditches for old roses. The daffodil people rabidly rescue, too. Since bulbs hide most of the year, daffodil hunters have to rely a bit more on technology, using GPS and community-shared computerized maps. (You can find many of their tales and tricks in *The Daffodil Journal*, a quarterly magazine published by the American Daffodil Society.) Sometimes, these daffodil freaks even operate in reverse, taking bulbs *back* to roadsides and old home sites to help re-create their romantic visions. Learn from group experiences and don't be dissuaded by the thought that all the good stuff is gone. Plant resilience and ephemeral nature mean that you can go to the same place on a different day and see totally different plants to scavenge. Whether you do it alone or in a group, beware of the plants you're dealing with: what might be a sweet, scraggly rose or single daisy in a field in the wild might be a monster in your garden's healthy soil.

A few years ago, while wandering around her little hometown, my mother took a few cuttings of a rose she spotted near the abandoned Brunson Laundry. As shown in chapter 5, Gloria is a great rooter; she's also a great collector. Against the laundry's brick wall, with terrible soil and no water, it was a little rose bush with tiny white flowers. Gloria rooted it and planted it right

in the vegetable garden. For the most part, Gloria prefers to collect in old gardens, where the owners can tell her a bit about their plants and share stories and cuttings. Otherwise, as with that tiny rose, you never really know what you're getting. Unfortunately, that rose soon revealed its inner monster, covering the entire fence where we normally grow cucumbers. We later identified it as 'Alberic Barbier,' which can reach 60 feet tall with trunks 6 inches around. Getting rid of that freebie cost us a few days' worth of prickly work. Worse are the things you collect and can never get rid off.

Seeds can be more difficult to deal with: even after you realize you've collected a weed, they come up for years. For a while, some friends and I collected the spectacular, fuchsia or red-flowered annual Flora's paintbrush (*Emilia coccinea*) in central Florida. We quickly grew dozens of them with pride, but by July, they were flowering bright fuchsia in every bed, in every sidewalk crack, and worse, in the pastel perennial border, totally ruining the color scheme. Since then, I've found *Emilia* in orange, red, and yellow, and I now don't care as much (or at all) about color schemes, so I grow them all. I've collected Carolina aster, saltbush, and camphorweed, which has the smelly scientific name of *Pluchea*, along with *Packera tomentosa*, a new favorite weedy aster with tiny hosta-like leaves, also known as butterweed. These are all invasive native plants that spread aggressively by seed. If you don't realize this in time, you might as well get used to pulling up seedlings because these spectacular wildflowers make millions of seeds that will never stop coming up in the garden.

Flora's paintbrush thrives in harsh environs throughout the tropics. It will seed itself in and is easy to share.

With these caveats in mind, scavenging still thrills and builds intriguing, personal gardens. It doesn't have to be complex, time consuming, or even dirty. And if you don't have time or desire for any of this fieldwork, you can stop by and scavenge at your local grocery store. The florist section has been a good source of chrysanthemums for me. I buy the cut flowers, root them, and plant them in the garden. They don't all work out, though; many neon-colored mums that you see are actually white mums that are spray painted. And I don't mean lightly spray painted, as you might have seen a florist do, but spray-on-tan painted, so they appear uniformly bright yellow, pink, or green. You can scratch that off to reveal white-petaled mums,

and some, especially the big, floppy spider types, will grow outside, but they get mildew and don't make great garden plants. Another caveat is that cut flower stems are totally desiccated, so they won't root easily. I always cut away the ends or soak the entire cutting in water for a few hours before I root them, using the middle, which is still most likely to be turgid and root-able. Lemongrass, rosemary, and basil will all root, too.

In the fall, the grocery store is also my go-to source for elephant garlic, one of my favorite, stately, and useful garden bulbs. In the South, we can't grow giant, brilliant purple alliums like 'Lucille Ball,' but elephant garlic is a great substitute. In gardens, I use this reliable, perennial allium mixed into flower borders. In containers, I mix it with narcissus or pansies or seed in a pack of radish. Once, in need of thousands of bulbs, I wound up buying from a restaurant supplier. Planted in fall, thick, steel gray leaves emerge for winter. In mid-May, giant, bouncy, lavender flowers are topped with a tiny jester's cap. Still, almost no bulb companies sell elephant garlic. You might find plants adapted to your climate are best discovered on the produce aisle.

Potatoes are another great underground crop that you can get from your grocery store. In our climate, we plant potatoes around Valentine's Day when our soil temperatures start to reach 50ºF, which is the optimum conditions for starting potatoes. Pick up the cute new purple ones, golden fingerlings, and tiny reds, and cut them in half or quarters, leaving one of the eyes (which is sort of a dark spot or sprout). Let them sit a few days and then plant them in a bed with your pansies. As the pansies fade in the heat, the potato vines will scramble around. They add a dark, elegant green to spring gardens, but they quickly fade or get chewed up by beetles. After that, it's time to dig them up and eat. There is, however, a caveat here, as well.

Remember when potatoes used to sprout in the bag? Now many things in the grocery are sprayed with growth retardants for longer shelf life. Those synthetic chemicals will keep plants from growing in your garden, too. It's not always necessary, but to be safe and insure successful gardening, I always scavenge from the organic section. The grocery store garden list can go on and on. Try things, test things, and discover the fun of grocery store scavenging for yourself.

You probably know the old tricks: place pineapple tops, avocado pits, or carrot tops in water and watch the magic—we've all done these things. But when I give presentations and ask how many have done these experiments with their grandchildren, few hands go up. We're forgetting to pass them down. And while they may do germination experiments in their science classes at school, they're usually not spending any time on old tricks like cutting a potato in half to sprout it in the cap of a mayonnaise jar turned upside down and filled with water (even if it sometimes turns into a powdery blue moldy mess). Not only are these great lessons to share with our youth, they're great sources of free plants.

In a sense, our nursery and farm was started with a scavenged plant. At 300 or 400 pounds, it was the largest clump of crinum that I've ever

Cool chrysanthemums can be easily rooted from old gardens, cemetery flowers, or the grocery store. Here, my mother gathers a bunch of flowers.

Bill Adams, who first introduced me to *Crinum* 'Pink Trumpet', roamed backroads and never hesitated to stop and ask for plants. Here he holds up healing squash.

RIGHT 'Pink Trumpet' crinum in a vase. This plant came to me as an unknown. Waiting, watching, and searching for the real name of a plant show respect for the original breeder.

collected. I'd watched it for years in a trash pile outside of a juke joint way out on a lonely road near Allendale, South Carolina. I could see that this flower was different from others in its group, and I finally asked permission to dig it. I didn't ask him for his tractor or his help, but the farmer got into the idea of saving this plant and helping to share it with this young fella who wanted it. With his tractor, chains, and my truck, we wrenched it out of the trash pile behind the bar once named Regina's Disco Lounge. Years later, after having crinum experts and mentors compare it and recognize its characteristic size and late flower season, I named it for the bar, and *Crinum* 'Regina's Disco Lounge' became the basis of our nursery.

On another occasion, walking through a rough part of Charleston, South Carolina, I stopped to photograph this gorgeous, salmon-tinted crinum growing to the side of an abandoned house. I desperately wanted a piece of it, but standing on a tight little street under the interstate, I could feel eyes tracking me and dogs barking at me, so I kept my camera close and didn't make a fool of myself by trying to dig. Just then a tall, thick twenty-something neighborhood ambassador slid off his porch, in pajama pants and flip flops, and asked in creole, "Wuh u doin'?" I told him I was just taking a picture of the lily, and he replied, "Ah gotta' luly. Wanna see my luly?" The only polite thing to do was to follow him to the back of the house to his pride, an enormous patch of red canna lilies. We talked about bug issues, and I took his picture and some canna roots he dug up for me. He later led me to an abandoned housing project where I found yet another crinum lily that I hadn't seen before.

Inside a demolition area, tractors lined up for a Monday razing, I saw a unique, pale pink crinum lily flowering in the shade of a live oak. Again, after digging there, those few bulbs became a mass planting, became rows in the field, got tested in various garden settings, and sent out to experts to name. It's a common hybrid, but an unusual variation that we now call

"Mo' Pon"; the nonsensical-sounding name came from the first place I did a mass planting of this spectacular shade bulb, under cypress trees along the muddy shore of a place called Moore Pond. The name is a tribute to the pond's owner, a southern gentleman whose accent turned those words into "Mo' Pon."

Getting plants and getting names are different things, though. In our nursery world, names matter. It would be unethical and unprofessional of me just to take a found plant and immediately give it a name. I spend years asking around before I put a name on something, which has saved me in the past. Once, Bill Adams, friend and fellow plant scavenger, brought a box of giant crinum bulbs to me. I grew what became the largest, tallest, most stately crinum I'd ever seen. It was a plant I wasn't familiar with, with flowers 7 inches long hanging from 5-foot-tall stems. As is always my first step in identifying something that I can't find in any books, I sent pictures to a colleague, who was able to name it right away. 'Pink Trumpet' was the creation of one of my crinum mentors, in south Georgia, who sold it locally at his garden center, hundreds of miles from me. I was glad I asked. At home, you don't really have to worry about proper names like we do in the nursery business.

Crinum asiaticum under the steps in Charleston, South Carolina.

I keep a stock of scavenged perennials, especially chrysanthemums, that I can use in gardens and landscapes. Clients like the stories that go with them, the connection to my mentors like Ruth, Bennett, and Gloria, and they like (and I like) plants that *used* to be cool. Plant trends, as with fashion, music, and technology, change more quickly these days. Plants come in and out of our lives more frequently. These changes used to be based more on improvements: better, shorter, pinker, and taller varieties would replace those that came before. Today, plant trends come and go based on marketing, promotions, and consumerism. And if you're always buying, you're going to have to eventually throw things away. Lots of plants are too good for that, no matter how sexy the ad for the new plant.

Crinum "Mo' Pon" is pale pink in part shade.

Jean Ann Van Krevelen's Tips for Scavenging Plants in Urban Settings

Jean Ann Van Krevelen is pretty hip; in cities across the country, she always seems to know the coolest people. On social networks, she'll tell you about the best scarves, stamped nail art, and newest plants for balcony gardens. She's written two books on growing healthy food and runs the Gardener to Farmer website with great advice for urban gardening. She tries hard to keep me hip; she's even turned me on to Stitch 'n Bitch knitting. Here's her take on scavenging in the city.

Most of us live in more urban (or suburban) areas, anywhere other than the swamps of South Carolina. But that doesn't mean you can't experience the joys of scavenging. Gathering plant materials in the midst of rush-hour traffic can be just as exciting (and dangerous) as wading knee deep in alligators!

Us urban collectors have a different set of rules; the mean streets of the city demand it. We employ a few guerilla warfare tactics, cause that's how we roll, yo. Before you head out to collect, make sure you are geared up for the adventure:

Travel light. Forget the shovel, we're talking a hand trowel, good pair of bypass pruners, and a small pair of snippers. While you are at it, forget the car, too. If you can take public transportation, you are doing the environment a solid.

Take a reusable bag. Grab a lightweight, easy-to-clean bag that slips over your shoulder. Bonus points if you can wear it across your body and free both hands.

Keep it moist. If you are taking public transportation, it might take you a while to get home. To ensure your cuttings make the journey, make a little kit using damp paper towels and a plastic bag or two. Wrap the cut end or roots of any plants you scavenge in the towels and add to the plastic bag. Bonus points if you reuse plastic bags that came with a purchase.

Know your neighborhood. Personal safety is your top priority. If you are unfamiliar with your city, take along a buddy that can keep you from heading into unsafe areas.

Don't be greedy. When you do find plants, it's preferable to take a cutting or two, harvest seeds, or look for seedlings. Keep in mind the doctor's oath: first, do no harm.

Now that we have the appropriate gear, let's talk about where to collect plants. In the city, we don't have a lot of wilderness harboring yet-to-be-discovered varieties. More likely, we have abandoned lots and buildings, sidewalk cracks, parking lots, and the like. Of course, we also have neighbors, public buildings, and city plantings. Yet all of these properties were or are owned by someone, so how do you know what is okay to take? I've created a list using stoplight technology to guide you in your collection efforts. Keep in mind that these are guidelines, not the final word, and certainly not legal advice. Know the laws in your community!

RED LIGHT	YELLOW LIGHT	GREEN LIGHT
Digging up plants from your neighbor's yard	Taking a single, non-damaging cutting from your neighbor's yard	Asking your neighbor for a cutting, division, seeds, etc.
Taking plants from places of business	Taking plants from beds of empty office buildings	Taking a cutting or gathering seeds from an empty office building
Taking all of the plants from an abandoned lot		Taking one or two plants, cuttings, or seeds from an abandoned lot
Taking plants from a city planting	Taking one or two non-damaging cuttings from a city planting	Taking seeds from a city planting (non-damaging)
Taking plants that are for sale at the store	Picking up pieces of plants that have fallen on the floor and are no longer a part of saleable merchandise	Asking the clerk if you can have said pieces of plants that have fallen to the floor

Pest Philosophy

*Taking a Holistic Approach
to Insects and Weeds*

*I*N THE PROFESSION OF HORTICULTURE, pest-control philosophy has
changed greatly since the widespread successes of Integrated Pest
Management (IPM), introduced in the late 1960s. IPM promotes min-
imization of problems through observation, early intervention, and less
toxic, disruptive methods. Though it took twenty years, nurseries, botanical
gardens, and classrooms have mostly adopted IPM and reduced—but not
eliminated—wasteful, rote, calendar-based spray programs. But in home
gardening, we've gone the wrong way, letting pesticides become more common and easier to use, while glossing over the environmental, health, and
financial costs. In some home gardens, seasonal weed and insect sprays have
become much more common than they were a few decades ago.

Misleading marketing and public policy lulled us into a comfortable
numbness. It can be all too easy to believe that there are people in our government, some benevolent "they," that wouldn't let things be sold that are

An old gardeners trick is to plant marigolds among other plants for protection from parasitic microorganisms in the soil. This simple pairing represents complex relationships in the soil.

really harmful. During the same time, pests have also changed. New pests have moved into new areas, and as we eliminate one pest, we create whole new problems. That was just the case with screwworm flies, parasites that used to kill warm-blooded animals; their larvae would bore into the animals' hearts. It was a huge problem for everyone who had a cow or goat, but the flies also attacked deer. In the 1950s, a program designed to save cattle and goats eradicated screwworm flies but unintentionally resulted in the explosion of the deer population throughout the country.

Our expectations changed, too. Magazines and social media now show digitally altered photos of gardens, encouraging us to long for and buy things that promise perfection, even though our hobby of gardening is often done to reconnect us with the natural and somewhat messy world. No matter when or where, pest control through domination is expensive. Our recent understanding of soil microbiology has only just begun to show

What is a pest? There's such beauty in the insect world and so many connections that we must learn to consider. Trying to kill one pest might kill something like these beautiful flatid bugs in Madagascar.

Beaver damage around the pond upstream from your house may indeed be a problem, but in the Congaree National Park in South Carolina, the beavers have stripped the bark off the trees to reveal the stunning pattern of borer holes.

us how damaging pest control is to beneficial bacteria, fungi, and insects belowground. We now have entire industries seeking ways to address the unanticipated consequences of pesticides.

To some degree with food crops, we've accepted that blemishes or bug holes are okay and may even indicate a more organic, toxin-free, safer fruit or vegetable. But when it comes to garden plants, we've gone the opposite way. We expect every plant to be blemish free; full of glossy leaves, symmetrically formed and already budded. Consider this change: just twenty-five years ago, you visit a nursery, where the family who ran it would work with you to help you pick out a good plant. Today, most people buy plants in a box store, from a sales clerk who rotates through the garden center, the Christmas-tree center, and the paint or lumber section. Every plant, therefore, must be tolerant of varied care and must have very broad appeal to attract a range of plant buyers. The nursery producing that plant aims for a pretty face. I think it's time for us, as gardeners, to remember that perfection in pots isn't our goal. It's time we recognize that a blemished rhododendron, like a tomato with a spot, may be a great plant that came from a producer who's watching and treating with pesticides only when needed.

All of this begs the question, what is a pest? From a holistic perspective, everything is a sort of pest to something else—even us—but some more so than others. So when do we decide that a pest requires some kind of intervention? My philosophy is that when a pest seriously threatens my family or myself, and the crops that support us, I intervene. But my first and daily responsibility is to make sure I do all I can to prevent problems before they get out of hand. Rust on crinum leaves, vole nibbles on the roots, and black spot on roses do not count. It's part of my job as a bulb farmer to help my clients and customers understand that plenty of healthy plants have pest damage. A bite out of a leaf or a dusting of sooty mold isn't much different from a liver spot on your skin. One of the grand dames of southern horticulture, Miss Hattie Watson, preached the live-and-let-live philosophy for over sixty-five years. Of most pests, including mildew on phlox, Miss Hattie recommends one of three options: pretend not to see it, pretend not to care, or go dig around your house to realize that you'll likely find more little whiskers of fungus on a pair of sneakers in the back of a closet, and focus on that. The takeaway here is that if you put things into a larger perspective, there's just no need for stress or toxic intervention in gardening.

———————————— The Teachers ————————————

MISS HATTIE WATSON AND KARI WHITLEY CROLLEY

Miss Hattie Watson and Kari Whitley Crolley are two experts who speak the same language. Miss Hattie's wisdom comes from years of intuitive experience and gardening at home; Kari's from science labs at the University of Georgia, breaking through stereotypes, and from running her own company. Both have roles as teachers, leading others to more successful gardens, and both espouse a commonsense, easy way of dealing with pests: live and let live.

Miss Hattie is the kind of person everybody in town thinks they know personally. She shared her garden wisdom weekly for forty years in a local newspaper column, through horticulture groups, and with her friends. She hosted garden open-garden days in Woodlanders Nursery, a world-renowned nursery in Aiken, South Carolina. She learned her craft from seventy years of dirty hands-on gardening: no schools, classes, or symposia. Miss Hattie once told me, "I had the bug. Always have, and you do, too. I can tell. There's just nothing to do about it is there? You do it because you're compelled to."

I also felt like I knew her in some deep, mystical kind of way well before the day we finally met in her room at Aiken's Lutheran Home. In that first meeting, she's sharp and intuitive, and we have a great time arguing over whether crinum foliage is too messy for the perennial border. She says, "It's just like a teenager's bedroom, you tidy it up one day, then . . ." the thought trailing off before she can finish. Organizing sentences wasn't easy for her that day; the effect of a small stroke frustrates her. Still, Miss Hattie is elegant and forthright and anxious to read a new gardening book. She keeps saying to another guest, "I *know* this boy," by which she means me. I don't think it's the stroke, this time; somehow, I know what she means. She knows me—my motivations, my loves, my passion for gardening, and my soul.

In talking, we discover that we have many connections—small-world, country stuff. She grew up in the swamp right next to where my grandmother grew up; she was a flower girl in my great aunt's wedding in 1932 or so; we both love the swamp plants. The names aren't coming to her, but I can guess her favorite ferns, vines, and spring ephemerals. When she says,

"Oh, the little white, comes up at Mother's Day," I know she means atamasco lily (*Zephyranthes atamasco*), which carpets the forest floor after spring floods.

I can quickly tell something else about her, too. Though she shares her wisdom and love of plants with every gardener in town that may have met her, Miss Hattie must have lived a somewhat reclusive life. I'm sure that lots of people would balk at my assessment; how could I say that about a local garden legend? But it's a feeling, and it's based on a few facts: she gardened enough to teach herself to do it like a pro; she made many beautiful sketches and watercolors of plants in her garden and in the woods; she wrote a lot, by hand and eloquently; and she loved the swamp and her famously jungle-like garden. These are all very solitary pursuits, and ones that I share.

I love a jungly garden, a place for plants. I want to see gardens where plants take control, get comfortable, let down their hair, lie across the walk-ways, and lean up against each other just being marvelous. I want gardens where the human touch is unseen—but not absent, of course; otherwise it wouldn't be a garden. Part of gardening this way means wrapping your arms and mind around everything, around the whole—and sometimes the whole includes bugs that suck and smell; hairy, sickly yellow funguses that squish between your toes; rodents that eat every root; even a stroke that keeps you from elegantly organizing.

There's no need for me to even ask; I know Miss Hattie gardened without trying to kill things. They were a part of her; she was a part of them. I'm not saying she was a Buddhist or anything; I'm sure she dealt with her share of pansy-eating rabbits and hosta-licking slugs. She even told me about how once she and her mother tried to trap an alligator. But when I asked, she couldn't name one chemical, one bottled pest control that she found necessary in her garden.

After I finish describing my father's tool shed, Miss Hattie tells me that she had one like it, too; full of brown, glass bottles with yellow, red, and black labels. Skulls and crossbones, sickening odors, and thick caustic liquids lined shelves in many tool sheds not too long ago. Dust and dusters, things borrowed from the local farmer, would sit for decades in my father's shed. They seemed like a good idea, but were too unpleasant to use more than once. Common sense kept him from throwing them away—he knew they'd do more damage in the trash—so they all just sat there, sort of safe, in glass bottles with disintegrating labels.

The first question most non-gardeners ask Miss Hattie is about what they can plant and where. But the second, unsurprisingly to me, is about

what they can use to kill a certain weed, bug, or mushroom. There is this urge in all of us to be in control. Today, pesticides are marketed to us as silver bullets: easy, bottled solutions that even smell nice. Cool labels promise butterflies and bounty. Sometimes they're really difficult to resist, but sometimes I go into a garden center and see shelves and shelves of bottled solutions and think, "When did gardening start requiring us to buy all these plastic bottles of stuff?" It doesn't. The best pest philosophy for gardening, for our health, the health of our plants, and for all the little creatures and connections that we'll never quite explain, is to live and let live.

Weeks later, I'm sitting with plant pathologist Kari Whitley Crolley, looking out over that swamp that Miss Hattie loves. Kari is a super-cool woman who runs her own Integrated Pest Management business, serving farms and wholesale nurseries all over the Southeast. She also serves as a board member of Charleston Horticultural Society and the South Carolina Nursery Association, and she teaches horticulture. I've seen Kari win over a conference hall full of really experienced landscape guys by issuing an opening

In gardens we make along waterways, such as this marshside garden, it's easy to understand how fertilizers and pesticides will run through the ground into the water. The same is true, while less easy to see, for any garden.

DDT and Malaria: The End Justifies the Means

In the 1940s, The National Malaria Eradication Program eliminated the disease in the United States by targeting mosquitos. Hundreds of thousands of homes, as well as swamps and fields, were treated with DDT, a pesticide used to kill insects. The program was based in Atlanta, Georgia, and focused on the South, but malaria killed people along the entire East Coast and in most of the Midwest. This eradication program was huge. The goal was to eliminate a widespread, fatal disease that lived in and was transferred by mosquitos. There are still a few older people around who remember when people died, right here in the United States, regularly of malaria. Massive applications of DDT solved the problem, but DDT also caused its own problems. For example, bald eagles nearly went extinct as residue from the repeated use of the pesticide washed into rivers, poisoning their food source and disrupting successful reproduction. Long-term effects of DDT are also linked to many health problems in humans. But it's important to recognize the difference between targeting a health issue and targeting home garden pests. Malaria still kills a million people a year, worldwide. Pesticide used to control malaria is a completely different thing from current trends of using chemicals to create mosquito-free yards. Mass application of pesticides through fogging, so that we can peacefully sit on the deck or play fetch with the dog, seems completely frivolous in comparison.

statement that makes them look around wide-eyed and poke each other like a bunch of schoolboys challenged by a girl: "I know more about pesticides than anybody else in this room." Kari has studied pesticides from the historical, chemical, human genetics, cultural, and even advertising standpoint. She's worked on nurseries that have used masses of pesticides and, subsequently, has deeply considered the impact on both her own body and the health of future generations. She knows the ways that pesticides work, how they kill, and she knows a lot about the unintended consequences of their use and misuse.

Kari asserts, "Pesticides totally, completely changed how we live. We couldn't live this way, in America, in any sense, without them." She means that in every way. She and I couldn't sit overlooking this marsh. We couldn't have had a glass of wine, while we picked some tomatoes from the garden. We couldn't turn on the tap for clean water to boil our pasta. We couldn't have the wheat used to make the pasta or the paper for the pasta box. We couldn't have "tamed" and populated this wild, buggy country.

The impact of pesticides is positive *and* negative: on the health of our water supply, insects, our crop pollination, our soil and rivers, and our bodies.

We want commerce, industry, and agriculture, which rely on pesticides, but we don't have a clue how those pesticides really affect us—the good ways are out of sight and mind; the bad ways may simply be unknown.

Kari's business is called Scout Horticultural Consulting. For homeowners, she'll investigate odd pest issues, look for underlying causes, and make recommendations for changes. She's like a crime-scene investigator for plant problems. I've watched her work, snooping around the base of a bay laurel hedge that for no obvious reason gets galls and canker on the stems. She takes samples and sends them off to labs, punches the screen on her phone, makes calls, and interviews everyone who might have taken a leak behind the hedge. A few days later, she presents a full-color, PDF report of all the issues that she has identified along with a few, commonsense solutions. The report begins:

> This is a very long report for a short diagnosis. I think that as a combination of initial improper planting depth and poor pruning, along with the hot, humid summers (not to forget the carbohydrate-robbing high night temperatures) and irregular cold patterns in the winter, these plants have been very stressed. I checked many galls and cankers for fungal spores. Even after seven days in a moisture chamber (to coax spore production), no pathogenic fungi were detected. Secondary (opportunistic) fungi were detected in some samples.

Even along rural waterways, homes, and golf courses, agriculture and industry nearby use pesticides and pollutants that flow into our rivers. The Edisto River in South Carolina is the longest undammed, blackwater river in the country. Here it looks pristine, but toxins in the water accumulate and poison the fish, as evidenced by signs warning against eating from the river.

You see how thorough she is? For home gardens, for giant nurseries, or for farmers, Kari investigates pest problems, scouts farms on a regular basis to preempt problems, and recommends pesticide and non-pesticide solutions. She has the right to boast:

> I tell all of my classes that I know more about pesticides than any one person they are going to meet. I know them from a huge, wholesale nursery perspective, from a farm perspective,

from a home landscape and garden perspective, from a biochemical perspective. I know the actions they take to kill things. And I tell them that based on that, I know that home gardeners have no need and no business using anything. That is a powerful statement.

I am thoroughly educated in regard to pesticides, and I don't own or use pesticides at my home. It's ironic that in an attempt to garden and be as close to nature as possible through the cultivation of the earth, that you would be willing to apply toxic pesticides to create your personal Eden. It shows how incredibly detached we've become from nature and how little we understand its complexity.

This is certainly a lot to ask of a modern home gardener, but it's the way Miss Hattie gardened. Some things die, other things thrive. It's a gardening philosophy that respects all life in the cycle. Even if you just stop using them, there are still enough pesticides in the world that our bodies and gardens are filled with them. Just like we don't use or need algaecides to add to a glass of water, we don't need pesticides to add to our garden. They are already there. There is no stopping the seep, through water, soil, and air, without huge societal and political changes. Even if you never buy one bottle, you bring the stuff in: when you buy a plant, when you buy a truckload of mulch, and when you buy a bale of hay. Lots of "ornamental" vegetables are sold potted in soil treated with a chemical that keeps them compact. These chemicals can have residual effects in the soil. When you remove that plant and put something else in, it too may be dwarfed. And be aware that plants sold as ornamentals, even if they are traditionally edible, may be treated with chemicals not allowed for food crops. A broader problem can come with mulch or compost containing chemicals.

Just a few years ago, uninvited pesticides killed hundreds of vegetable gardens across North America and Europe. It started with hay farmers spraying a new broadleaf pesticide on their grass fields. This pesticide targets non-grassy plants, so it can be applied over pastures. The pasture grasses absorb it without injury, while the weeds with broad, rounded leaves die. For government approval of this new product, the company said it would last in the soil about ninety days, then break down to inert chemicals. The EPA agreed. The treated pastures eventually became hay that got cut, baled, shipped as mulch, and fed to animals. According to how it was presented to the

EPA, that's where it should have ended. But things have unfolded very differently: animals passed it through, companies made compost from it, sold bags of that compost, and in a few years, vegetable gardens died. The residual life of that pesticide (aminopyralid), applied many miles away, transferred to those home gardens. Research from major universities has demonstrated that not only does the pesticide last up to seven years, it is transferred in cut hay to any gardener who uses that hay as mulch. Peas and peppers in backyard gardens shriveled in what became known as killer compost. This isn't an isolated problem; it's occurring all through the United States and in parts of Europe. The EPA is evaluating this, but meanwhile the pesticide is still in use. To buy hay for our crinum farm, we now have to find farmers, meet with them, and ask the right questions before we buy. For your home garden, look for a seal from the US Compost Council or use non-manure compost.

Pesticides come on the plants you buy, too. Almost every plant you buy from a nursery, unless it's a nursery that produces its own plants for on-site sales, has been drenched with an insecticide the day it leaves the nursery or has it incorporated into the soil, depending on the way each nursery is regulated.

More Specific Killers

As scientists work to find chemical pesticides that are more and more specifically targeted, they are coming up with better and safer solutions. Old-style, broad-spectrum pesticides might affect the grasshopper eating your corn, but they also hit bees, fish, dogs, and you because they target systems that all living animals have in common. A pesticide that targets a very specific part of a very specific pest is less likely to have unintended consequences. One of the promising fields of study is in insect growth regulators (IGRs). These are chemicals that mimic juvenile growth hormones in insects, altering the production of chitin (the compound insects use to make their exoskeleton) or by arresting development into adulthood. Some of these have become common in fire ant controls. They offer lots of promise as people-safe options. But if we've learned anything from the past, it's that we must proceed cautiously and slowly and evaluate everything through the scientific method and ethical considerations. The questions we don't know to ask are often the ones that show themselves as problems later. Do we really know that some pesticide that might kill by stopping the development of centipede exoskeleton won't do the same to shrimp? There are over 100,000 insect species in North America and only 1 percent are considered to be pests. When we do decide that we have to kill something that is causing us harm, we must be sure that there are no unintended consequences for related species.

Kaolin clay is mixed with water and sprayed on these grape leaves, forming a barrier. This is a sort of exclusions of pests that protects the grapes but doesn't harm the bees. Kaolin is an extremely fine-textured clay mined in South Carolina.

If you're using a landscape or lawn company, ask questions so that you know what they're using—or tell them what you want. Home gardeners account for less than 10 percent of all pesticide use in the United States. However, according to the EPA, that 10 percent means that we use about 66 million pounds each year. That huge number includes only what we buy off the shelves, and not the 90 million pounds that landscape companies apply to home and business landscapes. Some estimates break this down to about 6 pounds per acre—3 times the rate that farmers use. And that doesn't include those used in your park, the grocery store parking lot, the golf course, or the amusement park. If your neighbor, in the broadest sense of the word, is spraying, you're probably getting some of it. You cannot avoid the seep or the spray from down the street.

How did we get ourselves into this bizarre predicament where cow poop now has to be labeled as toxic? Where manure, spread on fields, gardens, and for renovation of woodlands, might kill garden and wild plants? Where pesticide residue is in our rivers, our food, and our bodies? It started with an agricultural industry that allowed itself to become too dependent on the massive application of chemicals provided by a few massive corporations. The horticulture trade has always followed and been influenced by agriculture. I'm a horticulturist from a family that's farmed along the same river since the 1750s. They and I see our professions as honorable; we have the knowledge and desire to use the earth to nourish people's bodies and souls. But the professions have changed. Individual cultivators quit asking questions and agricultural and horticultural societies quit meeting to discuss issues of the day. Agriculture affects every single aspect of life, and we need to be involved, to study, to discuss, and to educate ourselves and our children, especially in areas related to pesticide use, and to be a part of the decision-making process. This is equally important not only in agriculture, but also in the golf courses you walk across or the cut flowers you bring into your kitchen.

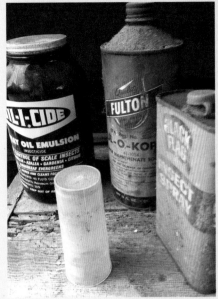

Old-style pesticide containers were less friendly, less promising than modern versions.

Ironically, I believe one of the causes for our current trust in the safety of pesticides results from the changes in chemicals brought about by Rachel Carson's classic environmentalist screed *Silent Spring*. When this marine biologist and conservationist pointed out the serious problems of the unregulated use of persistent, metal-based compounds, the public outcry forced regulatory changes. Chemicals were also changed drastically. Soon pesticides really were made more safely. We relaxed. Chemical buildups in streams and breast milk began to reduce. Surely, the new chemicals were okay? This is that lull I mentioned earlier; we began to assume that the pretty bottles for sale at the corner store couldn't be dangerous, simply because they were there. After all of this commotion, all of this debate, they just had to be safe. Didn't they?

Pesticides kill things. It's right there in the name; Kari likes to repeat a list of words—homicide, genocide, suicide, insecticide, fungicide, herbicide—to make the point that *-cide*s kill things, often indiscriminately. Besides that, we have an unimaginable number of chemicals now available, by some counts 400,000. Many are less damaging than those that we used to rely on, the pre–*Silent Spring* ones, but only *less* damaging. Many combine inside our bodies, in our soil and water, in ways that go unstudied. Even when there were no synthetic pesticides, pest treatments were dangerous.

Bordeaux mixture, used for centuries, is still promoted as an organic fungal control for late blight on potatoes, and it moves through the soil to accumulate in streams. It even has its own respiratory disease associated with it, "vineyard sprayer's lung." Rotenone, a common organic pest control, is highly dangerous to health and linked to Parkinson's disease. Pesticides kill.

But our recognition that pesticides kill has diminished. Modern marketing is a huge part of that, but we're all complicit for not asking enough questions. Remember those old chemical shelves in your parent's garage? Until the late 1960s, pesticides came in ominous bottles, with unpleasant smells and names. But many people back then were still tied to their rural backgrounds and understood the serious risk of pesticides and the importance of proper handling. Today, chemicals are too pretty, too easy, and too promising, and their instructions and warning labels are too small. We succumb to marketing. Someone understands our dream, or helps create that dream, and then sells us something to achieve it.

In reality, pesticides don't help achieve many of our dreams. The Worldwatch Institute indicates that for all crops, while pesticide use increased tenfold between the 1940s and the 1990s, plant losses due to pests also increased from a rate of 30 percent up to the current rate of 37 percent. In other words, things didn't get any better. If home gardening follows trends in agriculture, as it usually does, it's safe to say that in our gardens, there are no less pests, no more perfect gardens; in fact, there are likely no positive changes at all since we started using pesticides. If you're not sold on the rational arguments from people like Kari who've done the research, who've studied this, ask yourself: if pesticides do no discernable long-term good, why are we using them? Why are we gambling with our health, and the health of our planet? The difference between what we dream and what we run into outside too often becomes a frustration. We seek answers believing that someone has them. Every now and then they do, but the cost of the solution is too dear.

Updates and Adaptations

In the early 1980s, I arranged a booth representing the Clemson Horticulture Club. From it, we distributed apples and flyers diagraming pesticide accumulation in apples, which led to my being reprimanded by a professor from our department, even though I knew that this man understood that almost all apples contain pesticide residue. Since then, I've strayed in and out of thinking about and using all sorts of chemical pesticides. A few years ago, I finally quit them, at least in my own domain. On our crinum farm, we have complete control and I can be a bit of a purist when it comes to chemicals. Some people have countered that crinum don't have any serious pests, to which I respond, wasn't that a smart decision to grow them, then?

As gardeners and homeowners, we need to make good decisions like that. We need to quit wanting to grow the thing that is constantly covered in scale, for one example, but the answer isn't always quite that simple. Among our crinum, we grow our own food, too. We've made difficult choices. For example, we quit growing yellow squash because over the past few years, vine borers have made it impossible to grow without pesticides. In the meantime, we've discovered new squash, and we've learned how to cook with gourds.

For the larger populace, changing eating habits is surely more difficult and requires consideration and planning. But it can be done. As hard as it might be to believe, there was a time here in the United States when no one ate pasta. Once the Department of Agriculture decided that it wanted to help farmers sell more wheat, they came up with a plan to promote pasta consumption: macaroni and cheese! Our love of that American staple started with a government mar-

keting program. Nevertheless, changing food crops can be difficult. But Kari says changing garden and landscape crops should not be. From her years in nursery work, she's concluded, "We as an industry don't have any business selling plants that cannot thrive without the use of these products. If I had a plant that needed a fungicide in order to survive, I'd rip it out."

Sometimes it can be really difficult to make the call on how to deal with a pest problem. And although I do believe it's the same ethical choice— to kill any pest is to kill—it's more pronounced and dramatic with larger animals. Shoot a deer? Slam a vole with a croquet mallet? Trap a goose? My philosophy is that when we start interfering with nature, we have to deal with, or accept, the unintended consequences. We create pests by living in, developing, and interfering with a huge forest. We like what the forest provides, but we don't like to share it.

On our farm, our most serious mammal problem is adorable armadillos. We do our best to share with them, even though they present a serious threat to my family. As one plows through the crinum fields and yard, following the same path every night, he digs conical holes perfectly sized to break a person's ankle. With his nose in the ground, he rumbles between

Our crinum fields have been farmed for hundreds of years by lots of different people. Today we're moving back toward being a polyculture farm with varied crops and animals. Fennel is used to break up the soil, and it attracts a huge variety of beneficial insects.

Armadillos forage for insects in the nighttime and have quirky, fascinating lifestyles.

the pink lilies and the deep reds, crunching little grubs and mahogany-colored centipedes. His moonlit circuit returns him to a den with his brothers from dawn to dusk for sleep. Armadillos live in shallow burrows with friends of the same sex. Mothers will give birth to four or so identical young, either all brothers or sisters. After the young finish nursing, they move out to live in single-sex burrows. I think this is kind of sweet: fraternal, monkish.

But come daytime, I see the dangers of the armadillos' night foraging when my mother or nursery guests, walking through the fields, face the serious threat of twisting or breaking ankles. Those nocturnal fellas become "nuisance wildlife." I simply have to trap these guys, take them to somebody else's swampland, and release them. For a while, I used to shoot them, but one day, I did so, and he went under the house. For days, I could see his fresh trail of blood as he followed his nocturnal route, until finally I found him stiff in the road, and I began to cry over the sadness of it all. Exclusion seems like the best option to me. I've learned their habits and know that armadillos generally have a range that's respected by others. So we leave those that burrow out in the pasture alone, and harass any who come around the house or crinum fields. Our dog is actually our best harasser. It's sort of elitist, I know, and they probably bother the next farm down, but it's the best we can do.

We also deal with voles because our no-till management style encourages them. Fortunately, they are not into crinum so they're not a threat to our main crop. And, for whatever reason, they tend to stay away from our vegetables when we interplant them with the crinum. It could just be that they get their share, and I don't notice, or perhaps there is some associated resistance, where one plant sort of protects the other.

For the most part, though, crinum have few insect pests. Their long lives are a testament to their tenacity. The first crinum I encountered in my life had been planted in its spot fifty years before I found it and had not succumbed to bugs, fungus, or bush hog. My mother kept it growing for the past thirty years, and now it's eighty years old—that's a sustainable perennial. Thriving with little care, crinum lilies fit our management style and our philosophy of balance. We build our soil food web with compost tea, with the addition of fungi, with companion plants, and with crop rotation. Our belief is that healthy soil helps create a balance, preventing huge swings in any insect population. Most of these techniques have been discussed in previous chapters. A lot of what we do is experimental, and we can only give anecdotal reports on how things work. We test and trial new ideas, which are often old ideas reexamined. I don't know if our experimentation has broad implications outside of our fields, but it's a great opportunity to teach myself as well as visitors, and it sure gives me an understanding of how to use and help other people use the lessons taught by my mentors.

When crinum lilies, like other bulbs, do have problems, they're usually significant. Although there have only ever been a few viruses reported on crinum in the United States, I don't like to take any chances. If someone gives me a new crinum, I always isolate it for the first season. This is a step that home gardeners need to start taking, with found or shared plants, as well. In fact, there are several things that other knowledgeable growers and I do in

Pest Management with Mushrooms

Some of this stuff can really seem like science fiction: carpenter ants, who eat a bit of a leaf, may also get the spore of a fungus along with it. That spore grows, infecting the host and slowly shutting down its bodily functions, leaving the ant mummified. One fungus, after infecting a host carpenter ant, causes the ant to climb to the highest tree, where a mushroom pops out of the ant's head, killing the ant, but releasing thousands of spore to the wind. Managing these fungi, and using them as mycopesticides seems like a great idea. Mycologists, the scientists who study fungi, are just beginning to investigate the potential and risks of using fungi to control pests. Lots of technical and ethical challenges present themselves. First of all, ants and all animals, under constant attack from infection, have developed strategies to defend themselves against the fungi. A colony of ants or bees recognizes an infected comrade and may kill or eject him from the hive before the damage can spread. Mycologists look for ways around that. In doing so, this, like any pesticide, may become incredibly dangerous, and itself upset the ecological balance of the world. If we find a fungus in India that kills termites, how do we know that it doesn't also kill natural colonies of ants, beetles, or butterfly larvae? It's critical in any new science to ask questions, over and over, about the unintended consequences of a pesticide. It's also critical to ask who owns the intellectual and property rights to engineer these things? Who owns and should own the intellectual property associated with the science of mycology? This is biology. This is life. It belongs to the world. But this is technology where there's potential for huge profit. And that means there is the potential for management of mushrooms for pest control to turn into another corporate-political cabal. Or, even, at the risk of sounding alarmist, a mushroom arms race.

the field to prevent problems that might be good ideas for home gardeners to consider in general perennial care. We scout fields constantly, watching especially for insects that may cause problems—grasshoppers, mainly. In Florida and other consistently warm parts of the country, lubbers and other grasshoppers eat crinum leaves.

After the flowering season, we tend to let things go. The fields will get weedy, and the crinum leaves do get little rusty spots—in bad cases, we take those leaves away to keep the rust in check. Yes, it might look weedy, but to me it looks like a dynamic, well-balanced natural system where there is enough diversity that no one insect can sweep through or become a huge problem. That look is fine on our farm, and it's fine in our garden, too. For many of my clients' gardens, in late summer, when most perennials, including crinum, start to look tired, I cut them back to allow for new growth that will look great in the crisp days of fall.

Outside of my home, as a garden designer and manager, I can't always be a purist. I have a business of making people's gardens look the way they dream them. Occasionally I do turn to pesticides, but aside from soapy water and fire ant baits, I can't list one single insecticide that I've used in past

Plants and Pests from Around the World

For most bulb diseases, the only advice you'll ever hear is to kill your plants, put them in a plastic bag, and go plant someplace else; you'll never be able to replant bulbs on this land. Daffodils, tulips, and lilies—any bulbs—unlike most other plants, get shipped around the world. They've grown in various soils, with various pests. It is difficult to determine, since the bulbs are sold dehydrated, if they have pest problems. Most bulb growers would never intentionally sell infected bulbs, but it happens. The viruses, such as yellow stripe on the leaves of narcissus, will not kill off your other bulbs, but they can and will spread via aphids, making plants look sickly. Bulb farmers also struggle with nematodes and large bulb flies. To protect homeowners and farmers in the United States, the USDA established a preinspection program in the 1950s in the Netherlands, with an office with its own staff in Hillegom that is responsible for inspecting growing crops that are destined for the North American market. Healthy stocks are given an export permit for a particular lot number, and virus transmission in the stock is thus avoided. Bulbs also come in from India, Israel, and China, and all must have phytosanitary certificates provided by the USDA. But it's a huge industry with millions of bulbs to inspect, so you must be diligent, and be sure to buy only from reputable dealers. The concern and the harsh solutions apply to home gardeners and professional growers alike.

few years. No organics, no botanicals, and no insecticides. That change has been good for my body, my family, and the plants that we sell. In college, I had a job spraying camellias at resorts. The pesticide we used was an organo-phosphate, the kind of thing Rachel Carson wrote about in *Silent Spring*. Research shows that even low-level exposure to it can result in impaired memory and other health effects. I may have moved on, but they're still using it today; now there's some other young man or woman spraying the same chemical. And newer research shows that their children will likely have lower birth weights—all this for shiny leaves on a camellia bush. The same chemicals are used on lettuce, blueberries, and lots of other foods, which means you probably have this toxin stored in the fat in your body. But for that new generation of young gardeners or farmers, the exposure to these pesticides is heavy; it is an assault on their health and the health of their families. What happens to them will affect you, too.

This selection of red hot poker, *Kniphofia saramentosa* 'Riverbanks', was made for its late fall flowers.

In my design work, I sometimes use herbicides and weed killers, but I encourage clients to operate without them, and I design to reduce the use of them. But there are times when they're unavoidable. One garden that I managed bordered a cow pasture. When I started, it was literally filled with the most tenacious, perennial weeds. We discussed solarization and topsoil removal, but neither was a viable option. So with special herbicides, selected with the advice of a consultant, we spent two months target spraying. We understood the implications and problems we created by depleting the soil of its multiple microbial life forms, so to counteract that, we added a brand of organic, non-manure, woodchip-based compost in a top layer inoculated with soil mycorrhiza, and planted the entire garden with peas to help fix nitrogen and smother the remaining weeds for the remainder of the summer. We planted in the fall.

Still, the garden borders a farm fence, where deer passed every night on one side and a herd of Brangus cows on the other. There are no good options for keeping the deer out, though, so we planted an elegant and unusual mix of heat-loving deer-resistant plants to do the work for us. We even got away with some roses by completely surrounding them with a selection of pungent plants.

Perennials for an Elegant, Deer-Resistant Garden in Warm Climates

SCIENTIFIC NAME	COMMON NAME	MY FAVORITE VARIETIES	NOTES
Aspidistra	variegated cast-iron plant	'Spek-tacular'	evergreen groundcover for shade
Symphytotrichum oblongifolium	southern aster	'Fanny'	drought tolerant
Caryopteris divericata	chicken grease	'Snow Fairy', 'Pink Illusion'	interplant with roses to repel deer
Chrysanthemum	garden mum	'Miss Gloria's Thanksgiving Day', 'Button Yellow'	
Kniphofia saramentosa	winter-flowering poker plant	'Riverbanks', 'Valdosta'	evergreen; brilliant orange flowers in November and December
Salvia ×faranacia	blue salvia	'Henry Duelberg'	masses of flowers all summer
Salvia glabrescens	Japanese woodland salvia	'Mombana'	fuchsia flowers in late fall
Salvia nipponica	hardy salvia	'Fuji Snow'	yellow flowers, groundcover for light shade
Tagetes lemmonii	perennial marigold	'Compacta'	interplant with cosmos to repel deer
Teucrium fruticans	silver germander		can be sheered or loose

Finding the proper combination of plants required serious on-site testing and gardening with the understanding that, even with our planning, this garden might all be browsed on some hungry, icy night. The key is location. No national gardening program promoting plants resistant to any certain pest can be accurate. Plants and wildlife and their responses change with location. One of the commonly touted perennials for deer resistance in garden media is called bluebeard, *Caryopteris incana*, but that species shrivels in the Lowcountry of South Carolina. It hardly looks good for even one summer. Planting it here for deer resistance is a waste, but you'd be surprised how often it's recommended. However, *C. divericata* thrives, and deer won't eat its leaves, which are so pungent that I've always called it chicken grease plant—it smells like the backside of a fast food restaurant. This is just one example of how I look for pest-control solutions in my own gardens. When planning for pest resistance in your garden, seek the experience and wisdom of local botanical gardens and horticulturists. When dealing with any pest it is critical to understand that small changes in climate, soil, and even altitude change everything. A plant that works wonders in one location might have its own issues right down the road.

No matter the hype, most phlox get mildew. Gardenia and crepe myrtle, in humid climates, get sooty mold. If you really have to deal with these or other pests, note the problem, research it, and treat it at the right time. When we see sooty mold in late summer, it's generally too late in the year to address the problem. As Kari says, you need a professional who has the time and knowledge to understand the lifecycle of pests; and as Miss Hattie recommends, pretend not to care. Plant properly, ruthlessly remove plants that seem to attract pests, and consider that everything is alive and part of the overall system. Trying to control any one part disrupts the other. Let go of the desire to control and be amazed by the desire of plants to thrive.

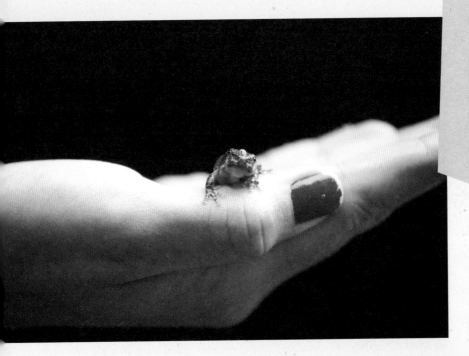

We have the responsibility of caring for and understanding how our actions affect others. Atrazine, a common herbicide used on lawns and crops that contaminates our drinking water, has been shown to cause demasculinization in male frogs.

The Buddhist Perspective

If you're the sort of person who will safely catch a roach, fly, or mosquito that's stuck inside your house and release it outside, you know there's a certain Buddhist quality to that action. It reveals a compassion for all living things. At the Charleston Tibetan Society and South Carolina Dharma Group, Tibetan Buddhist monk and spiritual director Geshe Dakpa Topgyal explains the ways Buddhists see all living things as worthy of our compassion:

For Buddhists, ethics means voluntary self-discipline, refraining from actions that cause pain and suffering in others or to yourself. It also means doing your best to help others, or, if that is not possible, at least doing no harm. This is a fundamental commitment, and we believe that good results come from good motivations and bad results from bad motivations.

To cultivate ethics, you need to understand how your actions and thought patterns make a difference in your own happiness, and the happiness of others. This is a unique potential of the human mind, heart, and brain. It is extremely important to investigate the connection between what you do and how it affects others. If an action would help you, but cause others to suffer, you must not do it.

Science and technology must not violate the laws of nature and the laws of cause and effect, or karma—that inevitably leads to disaster. Natural law is harmonious, creating the physical world, its structure and all the living beings who enjoy the natural world. Altering natural processes creates disharmony and imbalance and the effects are eventually always negative.

Interdependent, peaceful coexistence is the core of the fundamental Buddhist view on who we are, how we survive, what the outside physical world really is, and how it should be taken care of. The earth acts like a mother to all of us. Like children, we are entirely dependent on her. Taking good care of the world and our planet is like looking after our own home for our comfortable living.

Healthy and peaceful survival of all living beings including ourselves is entirely dependent on a healthy and clean environment such as: plants, grasses, flowers, soil, water, and air. Similarly, a clean and healthy environment is entirely dependent on healthy and moral behaviors, actions, and respectful use of the environment.

The use of chemical fertilizers and pesticides on any growing plants or grasses for our convenience is extremely bad for the environment, and ethically wrong. The use of chemical substances directly cause harm to millions of insects and the environment. It indirectly affects us with no escape. The latent force of chemicals and pesticides is much more seriously destructive than their immediate harm. We need a good-looking garden for our pleasures and comfort but surely not through harming other living beings and the environment which is a universal treasure belonging to no individual.

As human beings, each of us has the moral and ethical responsibility to protect the natural world. All living beings have an equal right to live in a healthy environment.

Geshe Topgyal also offers up this prayer:

May all living beings be free from pain.
May all living beings meet with joy and happiness.
May I have the courage of compassion and loving concern,
To look after the well being of all living things.
The pain and suffering in the world comes from cherishing oneself.
The joy and happiness in the world comes from cherishing others.
May my heart be imbued with love and compassion,
And able to replace self-cherishing with the cherishing others.
No one wishes even the slightest pain,
Or is ever content with happiness they have.
There is no difference between myself and others.
By realizing this, may I be able to invest all living beings in the ocean of joy.
MANTRA:
Na Ma Tsendra bhenza Trota Ya!
Hulu Hulu Titra Titra!
Bhenda Bhenda Hanan Hanan!
Amrita Hung Phed Sva Ha!!!!

Geshe Topgyal suggests reciting this prayer seven times and imagining that all contamination in the environment is purified through the force of this cleansing mantra.

Finding the Spirit

Telling Stories Through Your Garden

WHEN YOU LOOK AT ANY GARDEN, whether you realize it or not, you're looking into the world of the people who made and care for that garden. Some of the clues gardeners leave are intentional: placing a modernist sculpture in a historical garden, hanging ceramic luminary pumpkins, using crushed oyster shell walkways, or "planting" a sparkly bottle tree—each tells you a story about that particular gardener. If you look more closely, you discern other things about the gardeners, their culture, taste, and history. Every decision tells a story: an apple tree with a clematis trained up it, the yew hedge sheared tightly, hoses left out on the path, and even the level of the walkways. You can read a garden, as you do a house, a painting, or fashion. Besides design choices, in every garden there's a story of the biology of the place: rocks, wind, sun, mud, soil pathogens, and deer. Gardens are the story of two things: culture and biology. Gardeners are the storytellers.

A path through fields that have been planted with giant roses and small trees.

Gardening, for me, is a means of sharing an intense connection with the earth, of immersing myself in the dirt; it's a creative process. Sometimes that might take an afternoon, sometimes a decade. Sometimes it's just for me, for the pleasure of being in that moment, but usually I'm driven to tell a story. The story I'm most compelled to tell is the one that I've been telling in this book: that we are all part of a much larger process and in it together with a massive team of amazing creatures. It's all there for us to experience: smell this flower; take off your shoes—experience this garden barefoot; look what this frog does at night—let's see him under a blacklight; chew on this leaf; listen, that plant is whistling. Dirt, plants, and flowers are all sexual energy in a living mess of sensual pleasure. We're innately drawn to it; how can we not be? We are it. We are flowers, fungi, bulbs, and birds.

I am attracted to gardens that tell stories and reveal the spirit of the gardener and the place. I want to see into the gardener's soul. Though

LEFT This garden cascades through four different levels. Weeping plants tumble down the banks with the spirit of the land.

RIGHT This was a construction site just two years before this photo. Old-fashioned annuals (*Senna alata*, center) as well as antique milk jugs and gourd bird houses capture the spirit of rural farms.

sometimes it's not always the story that I want to hear. I'd rather see a garden that's telling a freaky story, like the Tuileries Garden in France, which has been the site of flagellations, massacres, and revolution, than a landscape where plants are used as a way to dominate and make things look neat and tidy, a static picture imposed on the earth. Sometimes I see landscapes in the slow process of becoming gardens. Sometimes I see a muddled picture created by a gardener who just doesn't have time to develop the story. People are often coming up to me at cocktail parties, or while pumping gas, or sending me emails saying, "I wanted this, but I have this. I wanted charming flowers, but I have a bunch of dead stuff. Or, I want to grow good stuff for my children to eat, but I'm not sure how." They ask me, "How can I make my garden?" It's a question I love to hear, because I can see that they're dreaming, and I have the skills to help them become a storyteller, too.

Once you get a handle on the basics of gardening, then you can focus on the dreaming, the fluffing, the telling of stories. Those stories need to be set up in the history of the site. I'm not talking about recreating anything, rather just recognizing what makes a certain place unique. Ask questions to find the spirit that comes from the natural history and the human impact of a place. Either or both may be majestic, sweet, harsh, or just plain boring, depending on what you learn. But that is the stage on which to tell our own stories.

The spirit of our old farm is in things like this set of wagon wheels that's been a favorite toy for children since the early 1970s.

I was fascinated with these kinds of stories as a boy. Leaning against a truck, old men visiting daddy would joke, "You boys know they planted that Magnolia Lane a hundred years ago so the women could walk house to the house in the shade? They didn't want the ladies to get tan back then." Even at eleven, I wasn't that gullible; a newly planted magnolia wouldn't shade any ladies. But I got to play in the trees, and I learned the lesson that plants are a part of culture, history, and myth. I did wonder and still do wonder at the visionary who planted them so many years before. How did he decide on magnolias? Who helped him dig them from the woods? And, most of all, did his "ladies" ever get to walk in their shade? Whatever the answer to these questions, his lane is a garden, and it captures the history of the land and its spirit.

The Magnolia Lane at Redcliffe Plantation, South Carolina, planted in the 1840s. The seeds from this eventually took over nearby eroded farmlands, creating a magical *Magnolia grandiflora* forest.

In a world where the artificial divide between nature and us grows larger every minute, we need gardens more than landscapes. We need gardens in public plazas, in parking lots and suburb entrances, in giant containers and in drainage ditches. We need gardens to tell our awesome stories, even if they are a bit chaotic—especially if they are mythical, emotional stories of how we are dependent on and compelled to be stewards of the spirit of every place.

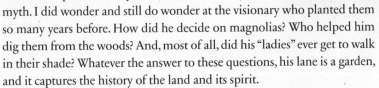

The Teachers

FRANCES PARKER AND DAVID HASKEL

Most gardeners possess some skills from both biology and art. I constantly seek advice from professionals in each field. In fact, sometimes I feel like I simply didn't have the courage to make a career decision between the two, so I became a horticulturist, which combines the best of both worlds. Frances Parker is an intuitive garden designer. She lives on the romantic, powerful, marshy land of the barrier islands of South Carolina. David Haskel is a biologist and a professor in the awe-inspiring mountains surrounding Sewanee, Tennessee. I sought advice from both of them on how to go about finding the spirit of place, finding what's crucial to making a garden that delights in the biology and the people that came before. Frances has been a mentor to me for decades, while David's recent book, *The Forest Unseen*, taught me decades worth of biology. They both point out that all of our five senses, and some sensitivities we can't easily explain, need to be engaged to make magical connections to the earth.

Frances Parker looking closely into her garden.

Frances Parker has the garden bug. She's a compulsive rooter and plant collector who's always saying things like, "I'm so happy to see those cuttings didn't root, I had nowhere to put them if they'd made it." Frances taught herself to design gardens when she and her husband, Milton, moved into an antebellum house in Beaufort, South Carolina. In the 1960s, theirs, like many of these once grand homes, had been divided into apartments. "We moved in because we needed a place, and Milton's parents had this old house. But, also, sort of to save the house; nobody wanted old houses then." Outside, the remnants of the original garden were still there because no one had bothered to get rid of them. This is a neighborhood infused with history. People wrote about it, drew it, and photographed it. The people who built it were rich, well educated, and into the cultural pursuits of the world's elite—including gardening. It was even once occupied by the Federal Army during the Civil War. For all these reasons, it's a well-documented place, so much so that Frances knows that one of the remnants in the garden, a bay hedge, was imported and plated by a Dr. Fuller in 1830. Leaving that intact, Frances amended, started indulging her plant passion, and pulling on farm roots; she grew most of their young family's food. Their garden's reputation grew, too. Soon people were asking if they could have their weddings in the garden. And many started to ask if she could design gardens for them. Frances kept gardening, and what started as her just sharing advice with friends and neighbors eventually grew into a full-fledged garden design and nursery business. But it never looked like a business: Frances Parker's headquarters and renowned nursery were in an old dairy barn on the back marsh.

The nursery was a ramble of little buildings off the beaten track. For plant lovers, it was a feverish place where you could always find one of these growing out of one of those; where you'd discover a treasure hidden behind an old pallet; where you'd be torn between exploring further and going back for a bigger wagon first. You could wander through paths of emerald potted palms, frizzy plants with leaves like silver threads, and purple orchids, and

From the street looking into Frances's garden, where you can immediately feel a balance of chaos and control.

Frances's nursery on the back marsh of the Beaufort River in South Carolina.

every once in a while on the ground you'd find a lemon yellow sphere that had fallen off some sort of citrus tree. There wasn't a business sign from the road to tell you where to stop, just a little pull off barely large enough for three cars, with some sort of concrete rubble to keep you from going in too far. A greenhouse of sorts was added on, and further down a path, you'd find the little cottage, the house where Milton was born in the 1930s. There were plants everywhere: tiny topiaries in clay pots, and gingers, ferns, palms, and amaryllis in plastic—some for sale, some for garden installations. Her collections were always mixed in, too: seed-grown coleus and antique rose cuttings from old farms and cemeteries, all arranged beautifully around a circular swept yard, under a live oak. Gray Spanish moss framed views to khaki cordgrass in the marsh. This nursery was a garden. Anyone riding by would look at the shack, see the plants all jumbled up and spilling everywhere, and wonder what cool old lady lived there.

When I first came to find Frances, I assumed this was her house, her garden, her style—maybe it was the home of someone who had to sell plants simply because they couldn't stop propagating. This was in the 1980s, before the Internet, before cellphones. A friend and I had set out on a road trip to seek out the infamously reclusive Frances Parker. You had to chase hands-on gardeners, nursery growers, and designers to find them in their gardens. We found Frances under that live oak. We became quick friends—she shared a load of plants, cuttings, and seeds all before 11:30 AM. It was hot out, and we'd been sitting in an old truck all morning and were tired—but Frances was just getting started. She had a cast on her left leg and a garden tour coming the next day. She said she'd buy us lunch and show us her other garden in town if we'd help her move potted lemons up the stairs.

So we left the little dairy cottage on the marsh, following her to the old part of town, locally called the Old Point. It's a little neck of land that juts out into the Beaufort River so that you feel like you're on an island. Early maps even label it "Beaufort Island." Antebellum doctors and planters built classically styled showhomes here. To get to Frances's town garden, we passed under a live oak limb with a "Clearance 9 feet" sign tacked up on it. An oxidized-orange Volvo wagon with moss growing on the bumper crowded the entry gate. Through a shaggy mock orange hedge, a grand white staircase descends into a parterre garden. This is when it struck me that Frances was more than a plant nut. She'd made two gardens: one like an old, poor woman's cottage garden, and one a highbrow, show-off garden—shack and chic. Over the decades since, I've realized that her gardens cover an even broader range of styles.

Boxwood hedges line brick paths, and a giant lawn fronts the elevated house. Parterres are filled with tropicalesque perennials, potted topiaries line the porch, and that original bay hedge makes a wall, hiding the magic within. For me, at such an early point in my gardening career, this was a real treat; I had only just met her, but I was going to have the chance to work in her amazing garden. Four hours later, my friend, a curmudgeonly type prone to feeling slighted, rolled his eyes and said, "She saw two men and thought she could get us to do her work. And all we got for lunch was a Bojangles' biscuit and unsweetened tea with a pale piece of lemon." But I was getting so much more. I learned how perennials should overflow the parterre, flop out, and cover the walkway. I learned how to thin them just enough that they look untouched but don't totally shade out the grass below. I came away with cuttings and an enduring friendship that later revealed some weird little 200-year-old connections with this land that's been gardened by farmers since 1810.

I didn't know at that time to ask the right questions: Who lived here before? How was this garden and land used in the past? But I've since learned that the answers to these questions can be crucial in your approach to designing your garden. And in the case of Frances's formal garden, I learned years later that the answers to these questions, which then only seemed to relate to the history of her garden, helped me to better understand part of my own family's history. From a cultural standpoint, those that have worked the land before us have left things behind that you have to work with—or consciously reject. Either choice reveals your mind and soul as a gardener. We can also look at this from a biology perspective, where those that came before might be rodents, fungi, giant live oaks, or tiny fiddler crabs. Their bodies, excrement, and lives make up the living layer, humus, and the plants. This is all your garden. There are so many questions to ask.

David Haskel reiterated this point to me recently. I asked him how we could see the garden unseen—by which I meant, how could we appreciate the biota? Given that he's a biologist, who takes pictures of slugs in their own liquid crystal worlds under rotten stumps, I expected a science teacher's answer about lichen, worms, or seeds that stick in between the split-toed hooves of deer. Instead he told me something that led to a question I should have asked Frances all those years ago:

We make connections by giving our attention to a place. That place does not have to be an exotic faraway land. It doesn't have to have an obvious history. . . . A complementary way of finding the uniqueness of place is to seek out stories of the natural and cultural worlds. What was here one hundred years ago? One thousand years ago? At the time of my birth? How have people used this land and how has this changed the place?

Where there are centuries of history, architecture, roads, and myths in a place, such as Frances's Beaufort, finding the spirit might seem easier. History and biology seem to swallow you up while walking in her neighborhood. Even walkways tell a story. A granite walkway looks out of place here. There are no rocks for miles. Time hasn't really softened the few places you do see granite paths, but bricks soften as moss grows over them. So Frances only uses brick. The same thinking drives her choices of trellises, fences, and even the layout of gardens. As we walk, Frances points to a new trellis that sticks out like a sore thumb. We discuss how the angles differ from the rooflines of the homes and even how the color isn't complementary to the existing architecture. Did that gardener choose to make a statement or did she just impose her modern trellis on this ancient landscape? Even without having studied architectural history, I can sense this. In a garden Frances is renovating, she's searched the country to match the original, twisted wire fencing that was used here in the 1930s. Understanding how things fit in or do not fit in is one of those sensibilities that help us feel the spirit of a place.

In the Old Point neighborhood of Beaufort, all the gardens and houses line shady lanes. The lanes and houses are designed to catch the summer breeze. Even from a few blocks in, you often have long views onto a wide, marshy river. The biology of this place is awesome. The light is beautiful and energizing. Humidity, warm days, and short rains encourage Spanish moss. Muck from the marsh was used by early cotton planters to fertilize their fields. All these things made this place a growing bonanza for early farmers. With this biology, and with the help of their enslaved brothers, they became the richest of American farmers. But for gardeners today, it does pose some challenges. We want to tell edited stories. Spanish moss–draped crepe myrtles are cool, but you have to constantly pick the stuff up and put it on the street to be taken away—for which you'll get red bug bites in the process. On full moons, the tide rolls in so far as to inundate front

Where there are no stones or rocks, it's difficult to use them in your garden.

Auldbrass Plantation

Of Frank Lloyd Wright's more than 1000 buildings in the United States, Auldbrass is his only one in the Lowcountry. It was planned as an extensive hobby farm, with a home, stables, aviary, and barns all inspired by the Spanish moss–draped live oaks and cypress of the blackwater swamps. It was only partially built, later partially burned, and on the edge of collapse when Directors Joel and Karen Silver embarked on not only saving it, but building out the unfinished plans. Like the live oaks, structures are angled and low. Like the moss, things drip and hang in balance. The farm's logo and maybe even some of the seating and table arrangements recognize the people who lived along these rivers for 30,000 years. The buildings are tucked into the landscape, coming in and out of focus among the trees, moss, and water. Eighty miles away, 1960s design work on Hilton Head Island's Sea Pines Plantation reflects the influence of Auldbrass. Landscape architect Robert Marvin revealed the spirit, chaos, and calm of the coastal forest by nestling homes and naturalistic gardens in the woods. In architecture, planning, and landscape design, this sort of celebration of the spirit of the Lowcountry took hold for a while. Auldbrass is rarely open, but will occasionally offer tours to benefit the Open Land Trust of Beaufort County, South Carolina.

yards with salt water. "See why I use pittosporums?" Frances says, pointing to a huge pittosporum tree growing right in the salty muck. Pittosporums will tolerate those times when the neighborhood is flooded. Wind also barrels through, dictating what will thrive along the narrow streets. Lots of this land is filled-in marsh, so you never really know what the dirt is, where it came from, or what you might find below.

On the outer islands, which are less protected from ocean winds than Beaufort is, new developers sometimes consciously consider the native landscape. Of the naturalistic gardens on Spring Island, Frances says, "I love trying to make a new house look like God just dropped it right in the middle of the woods and dunes, and nobody's planted a thing." Biology dictates. While past planters on the island saw domination of the biology as a way to make money, the new development values the biology; they use it as way to keep the place special, as well as a way to make money on real estate, attracting kayakers and wildlife visitors. Cultural values have shifted. Frances's naturalistic gardens tell a new story by using plants that have evolved to survive the harsh biology of the islands. Some people refer to this style as gardening by reduction rather than addition. That is, taking out a few "ugly" things to help people see the beauty of what is naturally there.

The spirits of this land and those who've lived on it for tens of thousands of years meet in a garden that Frances cares for and improves in the neighboring town of Yemassee. The sublime Auldbrass Plantation was designed by Frank Lloyd Wright, who left 500 pages of notes about how the place should develop over time, including small areas for gardens, which Frances has since developed. Wright's designs were inspired by dripping Spanish moss, leaning live oaks, and Indian arrowheads. Low-slung, cypress-clad buildings come into and out of view among the trees. Frances's garden there is inspired by that history—the notes, and designated garden spaces of the Wright design—and the possibilities of what she doesn't yet know that she can grow in the Lowcountry. Her gardening is inspired by Wright's designs, not planted according to any exact plan. I ask if she ever draws complete, detailed plans, and Frances says, "I don't do plans unless a contract requires it. How can you garden from a plan? There's always a giant root in the way or the light is funny here; you have to plan, then lay out on site. I do drawings after." This is gardening by the most pure dictates of biology.

As David Haskell says about the biology of a place, things that used to be there can be important references even in a brand new garden. Even if you are starting from bare ground, where topsoil has been scraped off and taken away, David advices that you look, listen, and try to get your head around what kind of habitat *wants* to be there. If everything has been scraped away, David says:

> An easy way to start is by poking around the "waste" and abandoned places nearby. Or, even better, natural areas such as parks. Often these places will be partly overrun with privet and other exuberant escapees from cultivation. They have their place, but they tend to get a bit out of hand and smother the other plants. So it takes some discernment to figure out the local plant community. An experimental approach is the most natural. Nobody can know what will grow best in a particular place. So, take a page from Darwin's book and allow many species to grow, then see which thrive. Those that do well will stay; those that wither will not. This sounds harsh and perhaps lazy, but I believe that planting a variety of plants and accepting that some will simply not make it is a big part of place-based gardening.

Smoothly shaped
pillars, planted with
jasmine and agave,
elicit a desire to touch.

In finding the spirit, start with that historical perspective, whether it is cultural history or biological history. Then start building your garden to engage the senses.

For most of us, finding the spirit of a place is visual. Gardening is about seeing colors, textures, contrast, and forms. We're drawn to the brilliant color of flower, or the vibrant yellow skin of a lemon that's fallen onto a mossy path. But these days, Frances pays more attention to texture and contrast; she loves greens. She says, "I don't look to flowers so much anymore. I've been through the matchy matchy and perennial gardening. It's fun, and the visual aspects are what we pay most attention to." But she achieves stunning impact with a completely green garden of 15-foot-tall, softly sculpted pillars of ebony green, fine-leaved jasmine, studded with pots of coarse agave, and crisscrossed by mossy brick walkways.

Another visual consideration is the play of light on a garden. How does the light change through the day? How do light and shade play out over the land and through the seasons? David wants to know this, too, when he's looking deeply: to find the unseen in a piece of land and see what the visual textures of the place are at different scales. He suggests starting down low and looking closely at the soil, say from a dog's perspective, then stand up and view it from your own, and finally get a "God's-eye view" using satellite imagery. "But," he says, "God dwells in the soil, so this metaphor is limited." The point is the visual world is often richly variegated. But vision is not our strongest sense.

David says, "After the eyes have had their feasts, move on to the ears—soundscapes reveal so much." How present is the hum of human activity? Is that a part of the garden, a part of the spirit of a place? Or is it something to be muted by garden design tricks? What bird or insect sounds are present? How does sound change throughout the day or throughout the week? The animal world makes itself known through sound; we hear thousands of crickets and birds, but see only a few.

We must care for the unseen animal world in order to have and enjoy their music. Pesticides to kill mosquitoes kill animals, too, but plants that provide food and shelter invite them in. When she wants to diminish the

unwelcome sounds, Frances relies on a trick of mind that she demonstrates by taking me on a walk around the Old Point to see gardens she's designed. It's 5:00 pm, and even in little Beaufort, there's traffic noise. At the end of the neighborhood, we enter a garden overlooking the swing bridge to Lady's Island. The garden is beautiful, the rocking chairs inviting, but the traffic noise is really loud. But I quickly forget all of that and focus on the bridge. "See, it's your focus," she says. "You see the marsh and bridge; you're enjoying the view; you focus and dismiss the sound." Away from the view, other little things distract you and even cover the sound. Standing by a small fountain, the simple trickle captures your attention. It doesn't drown out the traffic, but your focus shifts; you're listening automatically.

Brick works as a walkway that allows diversity to grow in the garden. It's important, as David Haskel says, to get down on your hands and knees to really see the garden.

The smell of a garden is as important to the spirit of the garden as sound. Frances drives this point home as we approach another garden through a long driveway. A sasanqua hedge separates the drive and back garden. It's a beautiful, new variety full of big pink flowers, but they lack the wonderful cinnamon smell of old sasanquas, a winter fragrance that is part of the Lowcountry's spirit. So Frances used a design trick to help people focus on the fragrance. At the place in the hedge where a walkway connects the driveway to house, she's replaced the pink sasanqua with two specimens of an old variety, nestled right into the hedge. The leaves, the plants, and the hedge look uniform, but the smell at the walkway is intense. The spot where you pass most often smells great and recalls the spirit of old-timey sasanquas. Frances smells everything. She pulls a weed, a camphor tree seedling: "When they mature, the leaves of course smell of camphor, but the seedling's leaves smell just like sassafras." She scrapes the skin of every hanging citrus and smells: "This one has a skunky smell." Then she tells me to squeeze a small, pale, plum-sized limequat. My hands are instantly coated in fragrant oil. She tells me to hold onto it, and I realized it's the perfect size and firmness to roll around in my hands, like a palm massager. It's such a tactile thing, resistant, squishy, and soft. There's nothing like holding that ball of sweet brilliant yellow. It's part of the spirit of this place.

I ask how many of these details she considers when designing a garden for a client. She says, "I consider all the details and share them with some clients who want to hear about them. Some want to discover them, and others never get it." As a designer, we sometimes play on and use these smells, these things that most people are not aware of, to capture the spirit of a place, whether consciously perceived or not. I love that cutting magnolia roots releases a rich sassafras smell. For fun, I sometimes make up a list, a chain of things that smell like other things: magnolia roots smell like crushed camphor seedlings, which smell like sassafras. I know that not many clients would ever follow this trail with their noses, but if I tell them about it, they love that the smells and associations are all there to discover.

David reminds me that smell can be used in a totally different way. Not so much for pleasure, but as a reminder of things we cannot see:

> We can't see most of the microbiota that rule the world (bacteria, protists, fungi), but we can smell many of them. The rich smell of healthy soil is the smell of plant-friendly bacteria. These smells change radically with the seasons and with the amount of moisture. Garden smells transcend the obvious delights of flowers. Leaf mulch, wet leaves, dusty late-summer bark on trees: these odors form a frame around the rest of our sensory experience.

Like sound, we can learn to focus on smells; we can learn which smells represent decomposition or problems in the soil. For example, sour, rank-smelling soil in a container or in the ground can indicate too much water and, therefore, problems of root rot. We can use smells to help us learn about the unseen biology of a garden.

Both Frances and David, with two very different approaches, are acutely aware of all their senses. They are also aware of other perceptions that help

The fragrant oils of 'Eustis' limequat released by rubbing the skin of the fruits.

them feel or see deeply into the land, calling on other processes not as easily understood. Sometimes these may be small signals like an old farmer's trick of understanding that sour grass growing in a field means the soil pH is low; or knowing that a rancid smell in the dirt indicates lack of oxygen that will inhibit root growth; or how to divine water for picking a great space for plantings. If I smell the spicy, molasses scent of magnolia roots when I dig a hole, I know to move somewhere else, as those robbing roots will stunt whatever I want to plant. Sometimes it just comes down to being more open to look and observe slowly and deeply with the intent of understanding what's going on between plants, soil, and other lives. Both Frances and David have spent a good portion of their lives learning to read between the lines and tell marvelous stories of the natural world.

——————— Updates and Adaptations ———————

Immerse yourself in this mystery. If there's one message I want people to come away with when I make a garden, it's that. Enjoy the mystery and get your hands dirty. Dreaming is one thing, but it's the doing that matters most. Try to let go of your end goal for a certain prescribed look. In public gardens, city parks, balconies, yards, and on my own farm, the process is most important.

I've been lucky enough in my life and career to make public gardens that lots and lots of people get to see, along with little patio gardens that might be enjoyed by just one person. In every one, the first thought is of the plants. What can plants do here? How can they be spectacular? How can they make a magical space? Walkways, containers, seat covers, and decorations come later. Their style is important to telling the garden's story, but their cost or trendiness is irrelevant. Gardens with spirit have everything to do with the natural resources of biology and creativity.

My ways of finding the spirit of a place pull together intense lessons from scientists in labs and equally important ones from old men in fishing boats. The basics of site analysis are science and procedure; these I can determine quickly as I walk over a bit of land. Some design tricks and understandings of how people respond to certain plants' smells and textures come automatically, too. For me, it's the artistic processes that have always been more challenging. Lots of people helped me nurture and embrace them. But I always, *always* try to challenge myself by forcing the question, how can I make this garden in a way that I wouldn't easily, automatically make it? Even though I seem to always end up with overflowing, layered spaces of curving lines, disorder, and wildness, I sometimes make myself start a design with the formal, symmetrical, and expected. I question my design, draw over it, and push it around on sheets of paper. But whether I'm relying on curvy lines or straight ones, the fine-tuning of a garden happens on the ground. Before I dig anything, I like to lay out lines, spaces, and even specific plants in a mockup, using spray paint, string, sand, and even bales of hay to mark where things might go. Then I sit back and "live" with my choices for a while and change things as needed.

By working on some big-budget, splashy garden projects, where the ordinary and extraordinary mix together, I hope to stimulate all kinds of people to make their own gardens with what they have. The British film-maker Derek Jarman made a garden of pulled-together, washed-up junk in the shadows of a nuclear power station in Kent. Pearl Fryar started his garden in Bishopville, South Carolina, by pruning discarded bushes to win the "yard of the month contest" and became an internationally rec-ognized topiary artist. Bennett Baxely started with a jumble of farm sheds and sprigs from shrubs pulled from the swamp. In both big gardens with budgets and little gardens with none, the spirit of the place is always the main ingredient.

On our little farm, where there's a lot of history to build on, finding the spirit might seem straightforward. It is, but sometimes embracing that spirit can be difficult. Our farm is a tad trashy, a little rusty and rough. The basic flow of it happened organically over centuries. Buildings, trees, and people came and went. For a while in the 1960s and 70s, the place was basi-cally abandoned. A grove of paper mulberry edged into the barns; a bank-sia rose literally engulfed the two-story smokehouse; Carolina cherry laurel took over the fig patch; and a jujube tree colony lined the drive, everything running a bit wild. We roped things in, and my parents enhanced the spirit of a sustenance farm. No tarting up was allowed.

We use the old stuff, but we also add new stuff—we layer things to keep the spirit. Our crinum packing shed is in an old chicken house. We've made a little classroom in an old woodshed with rebuilt mud and horse hair walls and a digital projector. A smoke house, car park, troughs, and now solar pan-els determine the flow and are the focal points of the meandering garden. Crinum and other crops extend in the same lines as the old vegetable gar-den. As a child, I was dismissive of daddy's vegetable style—a garden made with taut strings and tilled in super straight rows—but today, our nursery follows his rows. There's a certain irony to it, because as a garden designer, I'm constantly complaining about too much formality and straight lines. But our fields elicit the history of small sustenance farms.

Part of the spirit of this farm is that it's always been a gathering place. We cherish a 1900 photo of a group of African American men in the barren winter yard kicking back in kitchen chairs around a fire with a boiling kettle. A man who grew up here in the 1930s told me what a paradise for children it was—the halfway point on the dirt road where all his cousins lived. My parents made it a gathering place, too.

Today, serious plant lovers come for little gatherings and cocktails made from crinum tea. Old friends and young friends bring lawn chairs just to sit under the giant pecans. Garden clubs host their plant sales here. Gardeners are closer to the earth's spirit no matter where they garden, town or suburbs, but sitting on a quiet farm under big trees, over thriving soil, and refreshed by the air of fields and magnolia forest is an immersion that even the most avid gardeners love.

In places with big histories and spectacular nature, finding the spirit of gardens seems easy. Oceans, mountains, rivers, and deserts tell their own stories. Finding the spirit of and telling a story on abandoned agricultural land or blighted urban blocks is more difficult. Sometimes you look deeply and realize the story needs enhancing. I was once asked to make a garden in a huge, enclosed museum courtyard. The ruined urban soil had been the foundation for an old hospital and later became the red clay construction staging site. New bleak, beige walls soared around three sides, and on one end, a massive, old brick wall ended the view. The job: "Can you bring life to this? Can you plan a garden that celebrates the working yards of poor Southerners and at the same time is a place for elegant parties?" Instead of getting it from the land, we found the spirit in the museum's collections. The plants we selected share geographic origins with objects from the museum's collection. African crinum lilies connect with the tribal carvings and ceremonial masks of West African slaves. Cypress and tupelo trees connect with the Carolina rice planters' dugout canoes. In doing this, we're finding the spirit of all the people and plants that had been part of the area's history—we found the spirit of the place through diversity.

The bulb preparation area reflects the spirit and colors of the farm.

We are stewards of this diversity. Professionals, home gardeners, people who sit on tree-selection committees, and people who might become any of these have influence on how our world is planted. We have to select, plan, and plant things that thrive, that encourage more diversity without causing havoc. There are just too many great plants in the world for us to plant the same things over and over or to plant disease-prone monocultures. In some early American landscapes, the trees were diverse, in forests like canopies. Our forebears captured the spirit of that American forest in new American towns. But we've slowly moved toward monoculture: streets and parking lots are planted with just one type of tree, all genetically identical. Gardens of diversity are more resilient, survive threats of pests, and encourage other kinds of diversity. The mix makes for better gardens, gardens with more creativity and gardens that recognize the diversity of biology. David Haskell gives a mantra for gardeners, designers, and anyone choosing plants:

> Life thrives on diversity. So a botanically mixed collection of plants will attract a wider range of creatures than a collection of just one or a few plants. Most animals only make use of a small range of plants, so a wide taxonomic palate in the garden will result in a wider range of animals. This is probably more important than lushness.

Still, many of us desire that lushness; we work hard in our gardens to achieve it by adding lots of water and fertilizer. But we can learn to see beauty in the diversity and read the stories told through our gardens. David's a great teacher of this:

> I'm a big believer in exercises that open the senses. We tend to be hasty "lookers" and, thereby, miss so much. Our minds, also, tend toward dissatisfaction with the present moment, always yearning for something better in the future. The advertising industry does a good job of exploiting that yearning and never letting it settle into acceptance of what we have. There is nothing wrong with planning for a better future, but unless we can find existing beauty in the now, we're unlikely to be happy with the future.
>   We find wonder in unlikely places when we slow down and pay attention. This takes some practice. We have to decide to pay attention, then repeatedly return our attention to our place. The mind wanders—it is seldom content to remain in place, so we gently bring it back and literally come back to our senses.

It's so easy to let your mind race ahead in the garden, missing those chances to ask pertinent questions, to learn the story of a place. In the early part of my career, I paid attention to almost nothing but plants and gardening. I took long road trips in trucks outfitted for plant collecting, forged great friendships, built dramatic gardens, and even stimulated the garden economy. Surrounded by plants, seeking those plant people, I built entire relationships based in the garden. On that first trip to see Frances Parker and on dozens of trips since, she and I have talked only about plants. If we ate, we ate lemons or nuts while we walked in gardens. Our conversations were of seeds, nurseries, and new plants. I saw her husband only once or twice. I never asked about the history of her house. I wasn't asking the right questions, seeking the connections that David Haskel encouraged me to seek. Twenty years later, while writing this book, Frances's husband Milton brought out a small history book from which he read a history of their house that ended with, "Built in 1810 by a Henry Tudor Farmer"—my sixth great grandfather.

That's a cool connection. The fact that it took me twenty years to make it is a great reminder to slow down and try not to be such a hasty looker. Gardening can be so many things: industry, production, art, therapy, and grueling work. We need to wipe off our lenses so that we can see more clearly. In doing so, we can help others, with less time, less connection, to share in that focus. Look at this amazing world that we are a part of. We're part of the process, part of the dirt, and part of the spirit. Everything we do in the garden, we do to others, our children, and ourselves. We garden to share this story with those who might be focused on other things. That's our job as teachers, parents, and gardeners: put good stuff in and keep bad stuff out. We have a lot more to filter out these days. Remembering how we got to where we are today affects how we go forward. By focusing on the basic skills and the plants, we can meld the lessons of the old and lead ourselves to places we never knew existed.

My mother set the style and spirit of our farm. Over centuries of farm history, she wrote her own story by creating a place that is both inspiring and relaxing. She values the past, here holding her grandmother's milk jug, but she lives in the present, today using that jug as a vase for crinum from our mail-order nursery.

Old and Young Spirits of a Mountain Garden

I met Kelly Holderbrooks in her element, a garden near a peak of a Blue Ridge Mountain surrounded by spectacular rocks, moss, and fall leaves—it's wild enough to make you wonder if it is even gardened. Kelly and this garden lead us to new places and new types of gardening, encompassing old methods, nature, the future, and the spirit of the land. She and the garden fit perfectly here in the mountains, embracing, shining, and enhancing this fragile beauty so that we—and those that come after us—can enjoy the magic. Kelly told me:

I have always been drawn to playing, planting, and designing in the land.

When I began to settle into my adult life, I chose the region of the Southern Appalachians in Western North Carolina as my home. It was in these mountains that I discovered a passion for gardening and design. Soon after coming back, my mother said to me, "Let me tell you a story. Every year your grandmother and I would plant a garden together at her house. The year after you were born, I brought you with me. You were eleven months old and dressed in a bonnet and shorts. I placed you at the end of the row we were planting. I gave you a teaspoon, and you played in the soil all day. I should have known then that you would be working with the land when you grew up."

The shared experience and oral history of my family created the foundation for my journey as a landscape designer. I've navigated and developed my career through mentors, formal education, and most importantly lots of time spent in the land.

One of the most important aspects of design is identifying the spirit of the place or the genius loci. Each garden space has inherent elements, which are tangible and intangible. The spirit of the place falls into the intangible category for me. I have developed ways to interpret the land and recognize the spirit of the place. In doing so, I am able to design a garden space that stitches the past to the present. What follows is my method for discovering the spirit of the place.

Exposure:

- Visit the space often and spend as much time as possible in it, making sure to include visits to the space at different times of the day and seasons.
- Investigate the history of the land; this includes gathering oral history and researching natural history of the land. Explore.

- Use your five senses to experience nature through both a macro and a micro lens.
- Share your thoughts and experiences with others, preferably in the space. Resonate.
- Create space and time for your experiences to settle in your mind and body.
- Recognize your role as an agent of change for the land, and use integrity as a guiding principle. Ultimately, my method filters down to connection. In order to experience or identify the spirit of the place, I think a designer should seek to connect with the space and the land, through exposure, exploration, and resonation.

Kelly is director of programs at Southern Highlands Reserve, a non-profit organization dedicated to sustaining the natural ecosystems of the Blue Ridge Mountains. She's a landscape designer, an inspiring natural teacher, a woman who feels the land and is dedicated to passing on the traditions and forging a new vision of how we connect to, live with, and garden in this earth.

Resources

Tools, Sprinklers, and Supplies

Compost In My Shoe
Jim Martin
843-270-5804
www.compostinmyshoe.com
Organic vegetable gardening and community-supported agriculture and garden design services. Blog, videos, and garden design tips.

The Drip Store
980 Park Center, Ste. E
Vista, CA 92081
877-597-1669
www.dripirrigation.com
Watering and drip irrigation supplies for decks, gardens, or farms.

Hook Billed Gardening Knives
Tom Hall and Jenks Farmer
803-386-1866
www.jenksfarmer.com
Tools, books, seeds, and crinum lilies along with articles and photos of rare plants and special gardens from the South.

Hunter Desportes
Plant Database Design and Management
www.desportes.com
Plant collections specialist offering consulting services to gardens and municipalities. Nature photos and videos.

Machete Specialists
2101 West Broadway, Ste. 103
Columbia, MO, 65203
573-569-4562
www.machetespecialists.com
Machetes of any style and price made anywhere in the world. Instructions, safety videos, and all sorts of machete and knife accoutrements.

Mushroom Mountain
Tradd and Olga Cotter
129 Merritt Road
Liberty, SC 29657
864-855-2469
www.mushroommountain.com
Mushroom spores of southeastern fungi. Instructional videos, classes, and supplies for growing mushrooms in the ground, on logs, or in containers.

Red Pig Garden Tools
Rita and Bob Denman
12040 SE Revenue Road
Boring, OR 97009
503-663-9404
www.redpigtools.com
Hand-forged tools, reproductions of old gardening tools, and a selection of high-quality modern gardening tools.

Singing Hills Antiques
Gary and Jennie Wynaucht
417 West Avenue
North Augusta, SC 29841
803-441-8805
*Somewhere between a junk shop and
a fine antiques shop: old tools, sprinklers, yard art, and an eclectic collection of stuff.*

Seeds and Plants

Arrowhead Alpines
Bob and Brigitta Stewart
1310 North Gregory Road
Fowlerville, MI 48836
517-223-3581
www.arrowheadalpines.com
*Rare and unusual plants, many
grown from seed. Blogs, musings,
books, and such from plant hybridizer
Joseph Tychonievich.*

Felder Rushing
www.felderrushing.net
*Books, presentations, and photographs
of how people really garden, mostly, in
the South, but also from Felder's travels around the world. Links to Felder's
entertaining and informative radio
program, The Gestalt Gardener.*

Heavenly Seed LLC
Mike Watkins
206 North Fork Drive
Anderson, SC 29621
864-209-8283
www.heavenlyseeds.com
*Seeds of heirloom vegetables collected
around the South, including the seed
collections of Dr. David Bradshaw.*

Jenks Farmer, Horticulturist
Jenks Farmer and Tom Hall
www.jenksfarmer.com
803-386-1866
*Garden design and consulting for
homes, public spaces, and botanical
collections. Plant profiles, videos, and
tips for environmentally friendly gardening.*

J. L. Hudson, Seedsman
P.O. Box 337
La Honda, CA 94020
www.jlhudsonseeds.net
*Seeds of non-hybridized, beautiful
plants. A treasure trove of plants,
books, and essays about environmental issues.*

Naturescapes of Beaufort, SC
naturescapesofbeaufort.com
Coosaw Island, SC 29907
843-525-9454
*Southeastern coastal plain plants and
unique local selections. Botanical sur-
veys and ecological consulting, great
photos, and articles on southern wildlife and plants.*

Nurseries Caroliniana
22 Stephens Estate
North Augusta, SC 29860
803-279-2707
www.nurcar.com
*Unique plants, many collected and
imported from Japan, that are offered
nowhere else in the United States.*

PHS Specialty Narcissus
P.O. Box 6061
St. Louis, MO 63139-0061
*One of a dwindling number of
American specialty daffodil growers.
Historic and new hybrids.*

The Southern Bulb Co.
www.southernbulbs.com
*Pass-along bulbs, great videos, books,
and presentations on gardening in hot
climates.*

Woodlanders Nursery
1128 Colleton Avenue
Aiken, SC 29801
803-648-7522
www.woodlanders.net
*Southeastern native plants, rare
exotics, and local cultivars offered via
mail order and limited open days.*

Photography Credits

Chris Birkenshaw, page 52
Patrick Butler, page 144
Andy Cabe, page 108
Mike Creel, page 174
Hunter Desportes, pages 33 left, 122, 166, 167, 169, 170 right, 173, 188, 200
Tom Hall, pages 90, 170 left, 178 bottom
Will Hooker, pages 146, 147
Mark A. Lee of Great Exposures, page 59
Jim Martin, pages 84, 85
North Carolina State University / Rebecca Kirkland, page 143
Felder Rushing, pages 110, 111
David Schilling Photography, page 80
State Archives of Florida, Florida Memory, page 116
Virginia R. Weiler, pages 180 right, 187 left, 209, 210 right
Sharon Wilson, Scribble Time Photography, pages 53 top, 98 top, 225
John Wott, page 113

Public Gardens

Dunnaway Gardens, page 138 bottom
Jim Thompson House, Bangkok, page 140 top
J. C. Raulston Arboretum, page 143 bottom
Moore Farms Botanical Gardens, pages 91, 161, 162, 209, 210 left
Riverbanks Botanical Garden, pages 42, 43, 145
Redcliffe Plantation South Carolina Historical Site, page 211 bottom
University of Tennessee Botanical Gardens, page 45 left
Wat Xieng Thong Temple, Luange Prabang, Laos, page 144

All other photos by Augustus Jenkins Farmer.

Further Reading

Alexander, Rosemary. 2009. *The Essential Garden Design Workbook*. 2nd ed. Portland, OR: Timber Press.

Ban, Sue. 2009. *The Land That I Love: New Stories, Old Stories*. Self-published.

Blanchan, Neltje. 1909. *The American Flower Garden*. New York: Doubleday & Page.

Brown, Jane. 1986. *Gardens of a Golden Afternoon: The Story of a Partnership, Edward Lutyens and Getrude Jekyll*. New York: Penguin Books.

Chadwick, Alan. 2008. *Performance in the Garden: A Collection of Talks on Biodynamic French Intensive Horticulture*. Asheville, NC: Logosophi.

Creasy, Rosalind. 2010. *Edible Landscaping*. 2nd ed. San Francisco, CA: Sierra Club Books.

Del Tredici, Peter. 2004. "Neocreationism and the Illusion of Ecological Restoration." *Harvard Design Magazine*. Cambridge, MA.

Denckla, Tanya L. K. 2004. *The Gardener's A–Z Guide to Growing Organic Food*. Rev. ed. North Adams, MA: Storey.

Densmore, Frances. 1974. *How Indians Use Wild Plants for Food, Medicine and Craft*. New York: Dover Publications.

Edgar, Walter. 1998. *South Carolina: A History*. Columbia: University of South Carolina Press.

Free, Montague. 1957. *Plant Propagation in Pictures*. New York: Doubleday.

Fukuoka, Masanobu. 1978. *The One-Straw Revolution: An Introduction to Natural Farming*. New York: New York Review of Books.

Gainey, Ryan. 1993. *The Well Placed Weed: The Bountiful Garden of Ryan Gainey*. Indianapolis, IN: Taylor Trade Publishing.

Gaust, Drew Gilpin. 1982. *James Henry Hammond and the Old South*. Baton Rouge, LA: Louisiana State University Press.

Greenlee, John. 2009. *The American Meadow Garden: Creating a Natural Alternative to the Traditional Lawn*. Portland, OR: Timber Press.

Griffiths, Mark. 1995. *Index of Garden Plants: The New Royal Horticultural Society Dictionary*. New York: Macmillan.

Harrison, Mary. 2011. "Mary Gibson Henry, Plantswoman Extraordinaire." *Arnolida*. Boston, MA.

Haskell, David George. 2012. *The Forest Unseen: A Year's Watch in Nature*. New York: Penguin Group.

Hill, May Brawley. 1995. *Grandmother's Garden: The Old-Fashioned American Garden 1865–1915*. New York: Harry N. Abrams, Inc.

Holden, Edith. 1977. *The Country Diary of an Edwardian Lady*. Hammondsworth-Middlesex: Michael Joseph LTD.

Hillel, Daniel. 1992. *Out of the Earth: Civilization and the Life of the Soil*. Berkley: University of California Press.

Horn, David. 1988. *Ecological Appoach to Pest Management*. New York: Springer.

Hudson, Charles M. 1978. *Black Drink: A Native American Tea*. Athens: Univerisity of Georgia.

Jack, Zach, ed. 2005. *Black Earth and Ivory Tower: New American Essays from Farm and Classroom*. Columbia: University of South Carolina Press.

Jarman, Derek. 1996. *Derek Jarman's Garden*. New York. The Overlook Press.

Jeavons, John. 2012. *How to Grow More Vegetables*. New York: Crown Publications.

Knox, Gary, and Matthew Chappell. 2011. "Alternatives to Petroleum-Based Containers for the Nursery Industry." Univeristy of Florida IFAS Extension Papers. Gainesville, FL.

Kourik, Robert. 2009. *Drip Irrigation for Every Landscape and All Climates*. Occidental, CA: Metamorphoric Press.

Kramer, Jack. 1982. *Grow Your Own Plants*. Indianapolis, IN: New Century Publishing.

Lanza, Patricia. 1998. *Lasagna Gardening*. Emmaus, PA: Rodale Books.

Leighton, Ann. 1970. *Early American Gardens: "For Meate or Medicine"*. Boston, MA: Houghton Mifflin.

Lowenfels, Jeff. 2010. *Teaming with Microbes: Organic Gardener's Guide to the Soil Food Web*. Rev. ed. Portland, OR: Timber Press.

Lowenfels, Jeff. 2013. *Teaming with Nutrients: The Organic Gardener's Guide to Optimizing Plant Nutrition*. Portland, OR: Timber Press.

Maness, Harold S. 1986. *Forgotten Outpost: Fort Moore & Savannah Town, 1685–1765*. Self-published.

McDonough, William, and Michael Braungart. 2002. *Cradle to Cradle: Remaking the Way We Make Things*. New York: North Point Press.

Mollison, Bill. 1988. *Permaculture: A Designer's Manual*. Tyalgum, Australia: Tagari Publications.

Ogden, Scott. 2007. *Garden Bulbs for the South*. 2nd ed. Portland, OR: Timber Press. 2007.

Osler, Mirabel. 1990. *A Gentle Plea for Chaos: The Enchantment of Gardening.* New York: Simon and Schuster.

Peace, Tom. 2000. *Sunbelt Gardening: Success in Hot-Weather Climates.* Golden, CO: Fulcrum Publishing.

Peavy, William S. 1992. *Super Nutrition Gardening.* Garden City, NY: Avery Publishing.

Reichart, Peter, and Pathawee Khongkhunthian. 2007. *Spirit Houses of Thailand.* Bangkok, Thailand: White Lotus Company.

Ruiz-Lozano, J. M., R. Azcon, and M. Gomez. 1995. "Effects of Arbuscular-Mycorrhizal Glomus Species on Drought Tolerance: Pysiological and Nutritional Plant Responses." *Applied and Environmental Microbiology.* Washington, DC.

Rushing, Felder. 2011. *Slow Gardening: A No-Stress Philosophy for All Senses and Seasons.* White River Junction, VT: Chelsea Green Publishing.

Shein, Christopher, with Julie Thompson. *The Vegetable Gardener's Guide to Permaculture: Creating an Edible Ecosystem.* Portland, OR: Timber Press.

Slesin, Suzanne, et al. 1996. *Garden Tools: Everyday Things.* New York: Abbeville Press.

Stamets, Paul. 2005. *Mycelium Running: How Mushrooms Can Help Save the World.* Berkeley, CA: Ten Speed Press.

Talukder, Aminuzzaman. 2003. *Handbook for Home Gardening in Cambodia: The Complete Manual for Vegetable and Fruit Production.* Phnom Penh, Cambodia: Helen Keller International.

Watson, Hattie. 2011. *Hatties Columns: Gardening Articles and Observations by Hattie Watson.* Edgefield, SC: The Edgefield Advertiser Press.

White, David. 2000. "The Savannah River Site: Site Description, Land Use, and Management History." *Studies in Avian Biology.* Manhattan, KS.

Wiesinger, Chris, with Cherie Foster Colburn. 2011. *Heirloom Bulbs For Today.* Houston, TX: Bright Sky Press.

Young, James. 2009. *Seeds of Woody Plants in North America.* Rev. ed. Portland, OR: Timber Press.

Welch, William C., and Greg Grant. 2011. *Heirloom Gardening in the South: Yesterday's Pants for Today's Gardens.* College Park, TX: Texas A&M University Press.

Zobel, Anita. 2005. *Kolonihaver.* Copenhagen, Denmark: NYT Nordisk Forlag Arnold Busck.

Index

handmade structures. *See* trellises and sculptures

hardy salvia (*Salvia nipponica*)
  as deer repellant, 204
  'Fuji Snow', 204

Hartsville, South Carolina, 80

Haskel, David, 212, 216–217, 219–223
  mantra on diversity, 226
  on microbiota and smell, 222
  seeking connections and, 226, 227

haulm (perennial stems), 33–34

Henry, Mary, 174–175

Henry Foundation for Botanical Research, 175

Hill, Polly, 127–28

Holderbrooks, Kelly, 228–229

Hooker, Will, 143–144, 146–147

hummingbird bush (*Anisacanthus quadrifidus*), 105

Hunting Island, South Carolina, 113, 178

Huntree, Jane, 82

hyacinth bean, 144

hydrangea, 20
  rooting, 107

hydrozone gardening, 85

Indianapolis, Indiana, 60–62

indigo, rooting, 106

Indocalymus, 162

Integrated Pest Management (IPM), 186

interplanting, 36–38, 46–47, 49
  bean family for, 49
  mushrooms for, 65, 66–67

irrigation, 78
  drip system, 85, 86
  drip tape, 88
  emitter-style hose, 88–89
  germination problems and, 127
  history of, 78

plantation garden roses destroyed by irrigation system, 78

plant choice and eliminating need, 91, 93

plant selection and automated systems, 78, 81

regulated flow hoses, 88

soaker hose, 88

use of, wise, 88

Japanese woodland salvia (*Salvia glabrescens*)
  as deer repellant, 204
  'Mombana', 204

Japanese yew (*Podocarpus*), 48

Jarman, Derek, 224

jasmine, rooting, 106

J. C. Raulston Arboretum, 143

Jeckyl, Gertrude, 18

Jewels of Opar, 114

jicama, 144

jujube (*Ziziphus jujuba*), 224
  'Abbeville', 39
  'Li', 39

Kentucky coffee bean trees, 49

Keys, Andrew, 39

king stropharia mushrooms (*Stropharia rugosoannulata*), 66, 67, 75

kitchen gardens, 20

kiwifruit (*Actinidia*)
  as easy-care edible, 39
  'Ken's Red', 39

kiwi vine, 25

Knopf, Ruth, 98, 100–103, 162, 172, 183

kudzu, 64

lamb's quarters, 24

larkspur, 114

author's failed meadow and, 126–127

cobalt blue reversion, 124

letting go to seed, 124, 125

lemongrass, 180

lignin, 63

limequat 'Eustis', 221, 222

live oak, 85, 182, 215, 218, 219

loofah, 144

love grass, 72

LushLife Nursery (author's family farm), Columbia, South Carolina, 27–28
  "beyond organic", 33
  bulb preparation area, 225
  companion planting and, 30
  conversion of pasture to usable garden soil, no-till method, 69, 71
  crinum lilies of, 28, 29 (*see also* crinum lilies)
  crinum nursery at, 53
  crinum packing shed, 224
  database for, 166
  equipment used on, 163
  farm history, 34
  fertilizer used at, 53
  frost protection, 34
  green mulches, 35
  interplanting at, 33–34, 53
  irrigation system, 88–89
  mechanized blades used at, 162
  mulch materials, 34, 53, 54
  multiuse plants chosen for, 38
  no-till farming at, 54
  pest control at, 198–200, 202–203
  plant diversity at, 28, 53
  polyculture at, 199
  respect for life in soil, 35–36
  scavenged crinum lilies and, 181–183
  seed-in plants and, 125, 127

About the Author

HUNTER DESPORTES

Augustus Jenks Farmer, or Jenks, is a renaissance plantsman, with degrees from Clemson University and the University of Washington in science and botany, and a background in the arts from a family of artists, musicians, and farmers. As a garden designer, he has led teams to create and plant two of South Carolina's major botanical gardens, including Riverbanks Botanical Garden in Columbia. His designs for homes, museums, and businesses have received numerous awards and delighted hundreds of thousands of visitors. A natural teacher, he has established multiple internship programs and excels at getting people of all ages and from all walks of life to get outside and get their hands dirty. His pioneering mail-order nursery, jenksfarmer.com, specializes in organically grown, long-lived sub-tropical bulbs. He splits his time between a city garden in Columbia and his family's 18th-century South Carolina farm.